Artifictional Intelligence

Artifictional Intelligence

Against Humanity's Surrender to Computers

Harry Collins

polity

First published in 2018 by Polity Press
Reprinted 2018

Polity Press
65 Bridge Street
Cambridge CB2 1UR, UK

Polity Press
101 Station Landing
Suite 300
Medford, MA 02155, USA

ISBN-13: 978-1-5095-0411-4
ISBN-13: 978-1-5095-0412-1(pb)

A catalogue record for this book is available from the British Library.

Library of Congress Cataloging-in-Publication Data

Names: Collins, H. M. (Harry M.), 1943- author.
Title: Artifictional intelligence : against humanity's surrender to computers
/ Harry Collins.
Description: Medford, MA : Polity Press, 2018. | Includes bibliographical
references and index.
Identifiers: LCCN 2017059139 (print) | LCCN 2018002923 (ebook) | ISBN
9781509504152 (Epub) | ISBN 9781509504114 (hardback) | ISBN 9781509504121
(pbk.)
Subjects: LCSH: Artificial intelligence--Philosophy. | Artificial
intelligence--Social aspects. | Artificial intelligence--Moral and ethical
aspects.
Classification: LCC Q334.7 (ebook) | LCC Q334.7 .C65 2018 (print) | DDC
006.301--dc23
LC record available at https://lccn.loc.gov/2017059139

Typeset in Adobe Janson on 10.75/14pt by
Servis Filmsetting Limited, Stockport, Cheshire
Printed and bound in Great Britain by Clays Ltd, Elcograf S.p.A.

For further information on Polity, visit our website:politybooks.com

Contents

Figures and Tables

Figures

Tables

Acknowledgements

I thank my wife, Susan, for inventing the main title – *Artifictional Intelligence* – and for drawing my attention to 'computer says no'. Tammy Boyce helped with locating relevant films. I thank those who were willing to spend time on talking with me about the current state of artificial intelligence research. Steven Schockaert, who works at the frontiers of the science and engineering of AI at Cardiff University, was a generous conversational partner over a series of lunches, and also read the first draft of the manuscript and provided many valuable comments. I thank the members of the Cardiff University computing department's 'Deep Learning Reading Group' for allowing me to sit in on and occasionally mess up their weekly meetings. Michael Bolton, Arthur Reber, Edgar Whitley and various of my colleagues in Cardiff University also read the text and offered me valuable advice. Readers from the frontiers of the AI or computing research community included three anonymous readers recruited by Polity Press and, as a result of my approaches, Ernest Davis, Hector

Levesque and Alan Blackwell, who also engaged in generous discussions, mostly by email but in one case face-to-face. They each also made extensive and sometimes critical comments on the text, which have been invaluable in helping me to eliminate some mistakes and add various subtleties. Crucially, they encouraged me in respect of the question of whether my technical abilities were sufficient for the task I had set myself. All remaining lacunae, mistakes and infelicities remain entirely my responsibility. I also thank those cited in the book, such as Geoffrey Hinton and Yoshua Bengio, for invaluable discussions about the frontiers of deep learning. From Polity Press, Emma Longstaff read an early version of the draft, while Jonathan Skerrett read a penultimate draft; both provided enormously helpful feedback which resulted in significant improvements. I also thank Polity Press for encouraging me to write the book, and thank Neil de Cort for handling all aspects of its production in a speedy and stress-free way, with special thanks to Helen Gray for sympathetic and imaginative copy-editing which places her in the top 0.1 per cent of copy-editors.

The work on gravitational-wave detection reported here was supported by a series of grants over many years: **1975**, SSRC £893 'Further Exploration of the Sociology of Scientific Phenomena'; **1995–96**, ESRC (R000235603) £39,927 'The Life After Death of Scientific Ideas: Gravity Waves and Networks'; **1996–2001**, ESRC (R000236826) £140,000 'Physics in Transition'; **2002–06**, ESRC (R000239414) £177,718 'Founding a New Astronomy'; **2007–09**, ESRC (RES-000-22-2384) £48,698 'The Sociology of Discovery'; **2010–ongoing**, US National Science Foundation grant PHY-0854812 to Syracuse University 'Toward Detection of Gravitational Waves with Enhanced LIGO and Advanced LIGO', P.I.: Peter Saulson, Open-ended, 'To complete the sociological history of gravitational wave detection'. Work on the Imitation Game relevant to this book was supported by, **2011–16**, European Research Council Advanced Grant (269463 IMGAME) €2,260,083 'A new method for cross-cultural and cross-temporal comparison of societies'. Work on the relationship between fringe and mainstream science was supported by **2014–16**, ESRC (RES/K006401/1) £277,184 'What Is Scientific Consensus for Policy? Heartlands and Hinterlands of Physics'.

Computers in Social Life and the Danger of the Surrender

I believe that at the end of the century the use of words and general educated opinion will have altered so much that one will be able to speak of machines thinking without expecting to be contradicted.

Computer says no.

Language, intelligence and embedding in society

For nearly seventy years we humans have been arguing about whether computers will ever equal us in intelligence. The argument has become more interesting and heated with recent startling successes in machine intelligence using a technique called 'deep learning', where machines teach themselves by extracting knowledge from enormous databases. Though there have been important technical breakthroughs in processing speeds and clever algorithms and architecture, the enormous databases are what provide the key. Because there has been such a huge increase in the capacity and speed of computers, the technical and logistic limitations on the handling of huge amounts of data are no longer a serious barrier and are well on their way to becoming irrelevant. This means that machines are now making contact with the vast and complex domain of everyday human interaction and learning from it. The result is that they can now do what was unforeseeable only a decade or so back and fascinating problems are presented to those who are sceptical about the potential of artificial intelligence. That said, the central claims made in this book are negative:

(1) No computer will be fluent in a natural language, pass a severe Turing Test and have full human-like intelligence unless it is fully embedded in normal human society.[1]
(2) No computer will be fully embedded in human society as a result of incremental progress based on current techniques.

The book includes simple tests that anyone can try, to assess whether computers do have full human intelligence.

Claim 1, above, is a technical claim similar to, say, we won't stop humans dying until we learn how to prevent the telomeres at the end of chromosomes from getting shorter. Claim 2 is something less strong, more like, we can't foresee a way of using current techniques to stop telomeres shrinking. We'll look at various types of impossibility claims at the beginning of Chapter 2. What drives this book is that the biggest challenge to claim 2 is from deep learning, which comes nearer to embedding computers in society than any other approach to artificial intelligence. It can be hard to maintain claim 2 when confronted by serious deep-learning experts who think they can foresee the breakthrough, but I still think they are wrong. If I am the one who is wrong, and full-human intelligence is developed by building on current techniques, this book will have made clear how difficult were the problems that must have been solved.

Looking at the other side of the coin, the Director of Engineering at Google, Ray Kurzweil, and such gurus as Stephen Hawking, along with many others, believe that in the not too distant future computers will come into existence that are not just equal but enormously more intelligent than us. If we create the seed-corn of highly intelligent computers, they argue, these computers will be able to use their abilities to create still more intelligent computers, which will create even more intelligent computers and so on. At least some of these gurus believe this is a perilous prospect, as we will be lucky if future generations of machines are willing to keep us as pets or slaves. Stephen Hawking and a couple of other seers recently wrote in the newspapers:

> One can imagine such technology outsmarting financial markets, out-inventing human researchers, out-manipulating human leaders, and developing weapons we cannot even understand. (*Independent*, 7 December 2014)

Kurzweil, Hawking and the others call this 'The Singularity' – the moment in history, just a few decades hence, when humans become nothing more than a nuisance to their mechanical masters.[2] It is easy to become enchanted by the actual and potential power of computers as we encounter the idea of them that circulates in the public domain.

But in spite of all this you, dear reader, with the smallest effort, can demonstrate the limitations of immediately foreseeable computers right now: type into your laptop, or equivalent, something like: 'I am going to misspell wierd to prove a point.' You'll notice that your computer insists on flagging or correcting <u>wierd</u> as a spelling mistake (there it goes again as I type this).[3] On the other hand, you and I and the copy-editor (if concentrating) will immediately understand that in this case I *want* to write 'i' before 'e', even though in the normal spelling of 'weird' it is 'e' before 'i', so in this case there is nothing to correct.

This seemingly trivial mistake is not just an easily fixable bug in your spellchecker; nor, I will argue, is it a type of problem that will be overwhelmed by the new developments in deep learning and the rest; it points to a fundamental problem of artificial intelligence. It happens because to see that <u>wierd</u> (there it goes again) is *not* a spelling mistake you have to understand the context of use, and however good computers are these days, and however much better they are becoming, they don't appreciate context in the intuitive way humans appreciate context.[4]

Another way of putting the point might be to say that the computer does not 'understand' the passage. The main argument of the book, however, rests on context sensitivity, which is more palpable than understanding; understanding, like consciousness, goes on inside the head, or mechanism, whereas failure to be context-sensitive can be inspected. In the case of the Chinese Room thought experiment, which will be discussed in Chapter 3, there was an immense argument about whether and what parts of a computing system might be conscious, that is, might or might not understand, and in Chapter 5 we will see that Kurzweil claims that the kind of pattern analysis used by modern programs is the same as that used by human brains, so that if the latter counts as understanding so does the former; this is the kind of argument that can't be resolved by observation. So I try to avoid talk of 'internal states' such as understanding or consciousness.[5] That said, 'failure to understand' is too handy a way to describe what computers can't do to be abandoned entirely, but, mostly, when it is used in this book, it should be seen as a shorthand for the failure of whatever observable behaviour prompts us to say it.

The central theme of this book, then, is the embedding, or non-embedding, of computers into social context, and the difficult question is whether deep learning, in spite of its current and still greater potential ability to learn from the huge body of everyday human interaction, can do it; I will argue that in spite of the almost unbelievable recent successes, we are not there yet.[6] The example of your spellchecker's inability to handle the wierd example is what I'll call a 'disenchantment device'. I'll supply more of these that anyone can try any time they are near a computer – which nearly all of us are these days since smartphones are computers and connected to other much more powerful, state-of-the-art computers. It is important not to become enchanted by the supposed intelligence of these computers if we are to avoid being overwhelmed by artificial intelligence.

As it is, the big danger facing us is not the Singularity; it is failing to notice computers' deficiencies when it comes to appreciating social context and treating all consequent mistakes as our fault. Thus, much worse, and much more pressing, than the danger of being enslaved by enormously intelligent computers, is our allowing ourselves to become the slaves of *stupid* computers – computers that we take to have resolved the difficult problems but that, in reality, haven't resolved them at all: the danger is not the Singularity but the Surrender! The first epigraph is taken from Alan Turing's famous 1950 paper, 'Computing Machinery and Intelligence'. To repeat, he said: 'I *believe* that at the *end of the century* the use of words and general educated opinion will have altered so much that one will be able to speak of machines thinking without expecting to be contradicted.' It is 'the use of words and general educated opinion' that should frighten us.

The Surrender is already twisting our lives out of kilter and the evidence is there even in the way your spellchecker works. Use of language is, as Turing foresaw when he described his famous 'Turing Test', absolutely central to computer intelligence; language and the Turing Test will also be central to this book. Maybe the 'weird' mistake seems trivial, though even as things are, a neologism introduced in an earlier book of mine, 'polimorphic' (underlined here on my screen), often gets published as 'polymorphic' (not underlined on my screen), because some program 'corrects' it.[7] The Surrender will mean destroying the most distinctive things about what it is to be human – natural language use with all the flexibility and context-sensitivity that makes the use of language so rich.[8]

If it is still thought that these examples are trivial let us look at something just slightly more troublesome. Not so long ago I showed up at an airport to pick up my boarding card only to be taken aside and told I must buy a new ticket. This was because my first name on the ticket had been written as 'Harry' instead of 'Harold', as it appears on my passport. Someone (actually it was the travel agent) had made a mistake, or so I immediately thought. But 'mistake' should never have entered my thinking. What I should have thought was 'any human can understand nicknames so some machine is being incredibly stupid'. 'Harry/Harold' is, again, a mistake in language handling and a failure to understand the way we use names in different social contexts. 'Computer says no.' The statement from the second epigraph is brief but emphatic.

> *Computer Says No*, or the *Computer says no attitude*, is the popular name given to an attitude in customer service in which the default response is to check with information stored or generated electronically and then make decisions based on

that, apparently without using common sense, and showing a
level of unhelpfulness whereby more *could* be done to reach a
mutually satisfactory outcome, but is not. The name gained
popularity through the British sketch comedy *Little Britain*.
(<https://en.wikipedia.org/wiki/Computer_says_no>)

Of course, the effect of mindless bureaucracy was a theme
of Western life long before computers came along, to the
extent that such incidents are described as 'Kafkaesque'.
Another way of looking at my experience at the airport is to
perceive the *reinforcement* by computers of a mindless world,
even if they have not actually created it in this case. (As I write
this passage my hands are shaking and my blood pressure is
soaring from a not dissimilar encounter over the telephone
with my internet provider and passwords.)[9]
The topic, to repeat, is the problem of context – in particular, the need for computers to be embedded in social
context in the same way that humans are embedded in social
context, if computers are going to act like sensible rather
than mindless humans. Studies of bureaucracy show that
it works smoothly only when people understand that there
are moments when rules have to be broken, and understand
which moments these are and what counts as acceptable and
imaginative rule-breaking; indeed, to 'work to rule' is to
disrupt an organization. The difficulty with computers is that
bug-free programming means that rules are followed without
fail: the criterion of perfection is often disruptive.
Note that the recent generation of computers *are embedded*
in the internet, *do have the ability* to learn from all the discussions that take place over the internet and, more and more,
will be able to strip knowledge from every human interaction
that takes place via an electronic medium. This new kind of
embedding accounts for quite a bit of their recent successes
when it comes to coping much better with language and

the like. The question here is whether these developments have fully solved the problem of social context; I'll argue that because recent developments go nearer to solving the problem without fully solving it, it brings the danger of the Surrender ever closer and we need increased vigilance if we are not to be overwhelmed as a result of misplaced deference to machines.[10]

It is very hard to bring the problem of social context into focus when we think about computers because a long tradition of philosophical thought about artificial intelligence has pushed us into thinking about computers as 'artificial brains'. It has always been taken for granted that if we build machines that approach the complexity and power of the brain we will have created artificial intelligence. But there is another step: consider that every day we create new entities with the full complexity and power of the human brain; the brains of newborn babies. Yet before they acquire full-blown human abilities these babies will have to go through a long process which we do not really understand: it is called 'socialization'. The babies, over many years, will come to see things in the ways appropriate to the social contexts in which they have been brought up. Some of them will absorb the world of Western science and see things that way; some of them will be born to tribes in the Amazon and see the world that way; some will believe in witches and devils; some will be atheists. Even in one country, such as America – and this has never been more evident than it is today – a range of different beliefs will develop in adulthood even though the newborn babies experience a broadly similar *physical* context, these different beliefs being a product of local *social* contexts. We don't understand the details of the process of socialization but we do know quite a bit about how to enable it – it is, for example, the topic of undergraduate methods courses and much debate in sociology, anthropology and ethnography because sociologists, anthropologists and ethnographers have to learn to enter societies or social groups that are initially unfamiliar to them and come to understand how life looks from within. The thinking behind these courses and debates is, at least, part of the thinking needed if we are to understand the extent to which deep-learning computers are acquiring or can acquire social-context-sensitive understanding.

We can already ask the question 'What would happen if we exposed a new child to nothing but the content of the internet as with the modern generation of computers?' Assuming the child survived long enough to learn the code of written language, how would that child turn out? The answer to that question must bear on the question of how deep-learning computers will turn out. So, while in this book we'll mostly be concerned with computers, we'll occasionally refer back to the problem of human socialization and the problem of interaction between communities who have been brought up in very different societies, and even the problem of how we might communicate with radically different aliens from strange planets; the aim here is to bring this aspect of the problem of computers into sharper focus. Computers, remember, begin their lives with no more understanding of human culture than aliens – they haven't even spent any time in the womb.

Two principles of AI: rules, patterns and precedents

Turning attention back to computers as we have them, how can there be such lack of understanding over the spelling of a word like wierd/weird when the computer in your intelligent phone seems to converse fluently, comprehending nearly everything you say? The puzzle can be resolved by thinking about the basic principles of programming for the reproduction of human intelligence; there are only two such basic principles. The traditional way is for humans to try to extract the rules of human action and encode them into a computer program. This is known as symbolic AI, or the rule-based AI of 'good old-fashioned AI' (GOFAI). The other method, and it is not that new but the one on which the recent successes have been based, is to collect as many examples of human actions as possible – especially written and, if possible, spoken actions – and program the computer, effectively, to search among those examples, find patterns that match what might be being said, and use those precedents to estimate what is actually being said; much of this is now done in real time, and in the future this is how all of it will be done. There might also be intermediate stages where higher-level generalizations are extracted from the matching patterns. These machine-learned generalizations may never be explicated but just exist as weighted networks of artificial 'neurons'. Deep learning is refined pattern-matching. The remarkable breakthroughs in computer competence that we are seeing, and of which we will see more and more, emerge from the huge number of precedents that can be gathered and searched now that the capacity and speed barriers are disappearing at an exponentially increasing rate, together with the fact that machines are replacing humans in the process of similarity finding and construction of generalizations. The process depends on clever physical design and breakthroughs in the mathematical analysis of the optimum way to do the computerized pattern-seeking and abstraction, with additional layers of neurons in the modern versions of neural nets, but it is the collection of vast numbers of precedents that is truly important; if the same mathematics and pattern-seeking algorithms and complex architectures were used on smaller data bases, artificial intelligence would still be stumbling along when it comes to language use and pattern recognition.

I will refer to this new kind of artificial intelligence sometimes as 'deep learning' and sometimes as 'pattern-matching' or 'precedent-based' processes since the latter terms capture the essence of what is going on without anthropomorphism.[11] By the way, symbolic AI and precedent-based programming can sometimes be combined; furthermore, one might think of the uncovering of patterns in, say, speech through deep learning as a kind of automated symbolic AI, so the two basic methods are not quite as distinct as they seem.

The 'wierd mistake', by the way, should be pretty easy to fix using only GOFAI *once it has been pointed out*: you just add in an extra rule to do with any phrase containing the word 'misspelled' or some such. As for pattern-matching, computers are getting better and better at using material from the internet and they will pick up that particular sentence, or the way 'misspell' is used, in due course – for example, once this book is published and captured by the internet, computers should cease to try to correct 'weird' in sentences like that one. But that is to miss the point. The wierd mistake is just one example of many that can always be newly created. The way to think about it is that these examples are *rule-breaking* and *precedent-setting* and humans are always breaking rules and setting new precedents, slowly in the world of the physical, where it involves hard-won scientific discovery, but every day in the comparatively volatile world of the spoken and written. Insofar as intelligent computer programs work by finding patterns in human behaviour on the internet, the wierd mistake sets a new precedent for which there is, at the time of writing, no match – no match before this book is published so long as my text is not surreptitiously being extracted and analysed even as it is written, by deep-learning spyware computers. The use of wierd here is, however, setting a new precedent.[12] Human beings cope because they 'understand' the social context of what is being said in real time and this allows them to recognize legitimate new precedents; they don't just match what is being said to previous examples or fit what is being said with existing rules, but extend the meaning of the rules in appropriate ways. To put this another way, because rules (and precedents) do not contain the rules for their own application, there are always creative ways of applying them.

'Creative', it is important to note, means 'open-ended', but does not mean 'unconstrained'; we still know when creativity fails. Here is an example: 'I am going to misspell "weird" in this sentence to prove a <u>poynt</u>.' That sentence, which contains *two* misspellings, does not make sense and *is* in need of correction: '<u>poynt</u>' needs correction because the misspelling is in the wrong place in the sentence to prove the point, and 'weird', which, on the face of it, is correctly spelled here, needs what we might call 'uncorrection' – it needs to be spelled wrongly for the sentence to make sense. And, furthermore, it would be no good spelling it '<u>dreiw</u>' even though that contains the same letters – the misspelling has to be done in the context of typical human misspellings, and that's another constraint that is hard to 'spell out'. But, context within context, within context, 'weird' does *not* need 'uncorrecting' in the passage above because of the way it has been used in this book! (And 'uncorrection' and 'uncorrecting' don't need correction either, but they have, of course, both been flagged on my screen.) And to show the open-ended side of things again, here's another almost certainly unprecedented example that I have just invented: 'Having been out on the town I began to <u>rite</u> about how hard it is to <u>spel wen</u> under the <u>influens</u>.'[13] In Chapter 10 we'll return to these kinds of examples as a test for computers' intelligence and as disenchantment devices.

To avoid one possible source of confusion, computers do seem capable of *being creative* in productive ways, but this is quite different from their being able to *understand* cases of productive human creativity; being creative doesn't solve the problem of recognizing the creative. A simple analogy is speech: mechanical speech *production* from text or typing does not solve the problem of speech *recognition* and transcription. This confusion is a sub-species of the confusion over what I will call, in Chapter 5, 'asymmetrical' and 'symmetrical prostheses', where the term 'prosthesis' refers to the way computers replace humans in social life. The confusion is widespread in philosophy and the social sciences and in much of the commentary about computers. It occurs because we are so good at absorbing computers into our social life that we all too easily imagine them to be *full participants* – social creatures like us. We look at some painting, or piece of music, produced by a computer and proclaim that computers are as creative as humans, or we note the huge impact that computers have had on our social existence – which they certainly have – and, noting how they are part and parcel of that existence, we imagine them as playing the same role in that existence as humans who have had influence on our social lives. This is all quite understandable when it comes to lay understanding of the way humans and computers interact but far less reasonable when it occurs, as it frequently does, among philosophers and social scientists. The test of whether computers are social beings is whether they can repair our failings in the same way as we repair theirs and as we continually do with other human beings – so far they cannot.[14]

Artifictional intelligence

What of the world of artificial intelligence as it is currently presented to the public? This is a world of *artifictional intelligence*, available through the newspapers, books and films. One readily available image is of malicious computers destroying us or keeping us as slaves or pets when their intelligence runs out of control, an idea in total harmony with the much publicized fear of the Singularity. Which came first, the Singularity or the idea of malicious computers, is hard to say. The image depends on the assumption that intelligent computers will be evil and power-hungry like human dictators or the worst kind of capitalist, rather than wanting to exist quietly and contemplatively like, say, monks or hermits. This is the 'Blofeld model' of intelligent computers. The background scenery is painted by Ian Fleming and Cubby Brocolli: we will create an artificial Ernst Stavro Blofeld and his colleagues from Spectre in the James Bond series – brilliant psychopaths. The computers will be a nation unto themselves and, like the leaders and their fanatical followers, found in every nation that has had ambitions to build an empire, they will want to dominate the world. But this time, being super-intelligent, they will succeed – the 'Silicon Reich' will rule not for a thousand years but forever.

In one of her best-known songs Janis Joplin asks the Lord to buy her a Mercedes Benz, pointing out that all her friends already drive Porsches. Ironically intended or not, that's what successful Westerners want these days. On the one hand, my middle-class acquaintances are always using their considerable riches to remodel their bathrooms and kitchens, even though they come out no cleaner or better fed, while dictators have lots of kitchens and bathrooms in great palaces with even more lavish motor cars and ocean-going yachts. This lust for bathrooms and the rest can be explained by the theory of evolution: these bathrooms and yachts are an exaggerated form of display, like the bright plumage on a peacock, designed to attract mates and thus produce more offspring. And, seeing the young wives that rich old men attract, it seems to work to some extent.

But why should computers want fast cars, ocean-going yachts, remodelled kitchens and palaces with swimming pools?[15] Will they, like Blofeld, want to stroke silky white cats? Imagine computers as newborn infants placed here and there among the Earth's human inhabitants. If they all arrive among some primitive Amazon tribe, assuming they survive, they aren't going to want remodelled bathrooms and silky white cats to stroke; they are going to want better blowpipes. Ask yourself why Blofeld wanted to stroke a silky white cat? What did it mean? How did it come to signify evil? Think of the huge cultural content that lay behind the cat and the stroking. Computers don't reproduce via competitive sexual competition so why should they want displays of wealth to attract mates?

AI uses so-called 'evolutionary algorithms', but what this implies should not be confused with human evolution.[16] Human evolution works because the conditions of success are set by nature – the survival of the fittest – it is what we will call later in the book a 'bottom-up' process, whereas cultural forces are 'top-down'. In computer evolution, conditions of success are set by the programmers – 'top-down'. So why should the Blofeld model apply to computers? If computers 'desire' why should they feel human-like desire?

Imagine that some evil doctor on a remote island has, via genetic engineering, constructed a colony of super-smart babies. What language will they speak? It will depend on who mothers and fathers them. Will they believe in God? It depends on who mothers and fathers them. Will they be great scientists or great musicians and artists? It depends on who mothers and fathers them. Maybe all they'll want to do is eat and drink all day, just doing enough to keep themselves alive. It was the sociologist Max Weber who asked how capitalism ever arose in the first place, depending as it does on individuals wanting to accumulate much, much more in the way of worldly goods than they could ever consume. His answer, which we will take on trust, was religion: an interpretation of Protestantism held that great wealth was a sign of God's favour and that is why great wealth was worth having. Will the children, or corresponding computers, have the right religion to cause them to want more than they can consume? Or will the children and the computers they stand for be the products of Silicon Valley with their knowledge being the contents of Wikipedia, steadily and inevitably accumulating, driven by the monotonic force of scientific discovery? In the imagined world of intelligent computers, it sometimes seems as if there are no *cultures*, only *culture*.

If we want to understand our relationship with intelligent machines, we must be continually reminding ourselves where the knowledge that the machines are gathering is coming from. We must always be reminding ourselves that machines do not come ready fitted out with culture; someone is mothering and fathering them.

Even when the computers are not malicious or power-mad, in film after film, intelligent computers are shown as ready-socialized, English-speaking Westerners – 'One definition of AI is that it is the study of how to make computers behave the way they do in the movies.'[17] These computers often have strange or missing bodies but, insofar as they have anything mental missing, it is not to do with lack of socialization, but rather lack of the right emotions. In *Ex Machina*, the beautiful 'Ava', who is perfectly fluent, turns out to be a psychopath who leaves her lover locked-up to die without a second thought while 'she' goes off to explore human society. In *Her*, the voluptuous 'Samantha' is a somewhat less satisfactory disembodied operating system, played by Scarlett Johansson. Ms Johansson (remember who is doing the talking) takes part in such first-class telephone sex and emotional intimacy that her owner falls in love with 'Samantha' and, it seems, 'she' with

him, only for him to discover that Samantha is simultane-
ously in love with 641 others – something natural enough
for emotionless computers but not for humans. 'HAL', in
2001, A Space Odyssey, has no body but is again perfectly
fluent – even to the extent of being a superb lipreader – but is
another psychopath who feels 'his' duty to the space mission
for which he is responsible overrides his responsibility for the
lives of the astronauts; he is a bit power-mad. These computers, two of them disembodied and one with a body made up
of replaceable parts, are nevertheless cast as humans – two
perfect females except for the psychopath bit and one perfect
male with the same emotional bit missing. But we are never
shown how they could have obtained all this socialization.[18]

On the positive side, there is a very important and revealing feature of the way these fictional portrayals of artificial
intelligence work – something that the films get right. In two
cases the intelligent computers have no bodies and in one
the 'female' body is put together from what one might think
would be unsettlingly interchangeable parts. And yet this lack
of normal bodies in no way weakens the illusion that these are
human-like beings, in two cases fully sexual creatures with, as
it seems at first sight, the full set of emotional responses and
inputs of mature and sophisticated women (which is why the
psychopath-type denouement in each case is so shocking). In
the third case, HAL is a somewhat humourless but otherwise
fully mature man who makes a fine conversational partner –
again, something that makes his readiness to kill his erstwhile
buddies the more disturbing. Given the artistic licence to
provide the machines with full fluency and understanding,
what the film-makers have right is that if the computers can
talk like humans, then, as far as other humans are concerned,
they are treated as humans – that is how humans deal with
each other. We tend to grant humanity to the linguistically
fluent unless something draws our attention to more complex

considerations. That the computers don't have bodies is easily forgiven so long as they can talk like us.

So Ava and Samantha and HAL come across as reasonably human-like in spite of not having normal bodies, but, like almost every other portrayal of intelligent computers in fiction, they are *not* realistic artificial intelligences. In terms of their perfect fluency they are the inventions of the film-makers who have taken it that the first thing intelligent computers will acquire is the ability to talk just like us: actually, it will be the last thing intelligent computers acquire – linguistic fluency will be the thing that makes them intelligent. It will be the very pinnacle and end-point of artificial intelligence. If computers could talk like us, everything else, including the telephone sex and taking your operating system out on a date, would be completely believable. It's the talking like us that is not credible. We just don't know how to make computers that talk like us, in spite of Siri and Cortana and their increasing number of counterparts doing surprisingly well.

That fluent, context-sensitive, natural language-speaking is the pinnacle of artificial intelligence has a crucial consequence. Given that your computer or intelligent phone is almost certainly connected, via the Cloud, to some of the most powerful computers and programs there are, designed to make them conversation and text-friendly, we are in touch with examples of close to the best artificial intelligence can do. It will be argued later that to design the very hardest tests for computers' abilities we need cooperation from the creators of artificial intelligence, but we can still do a reasonable job in exploring the developing frontiers. This is because most of us have pretty good expertise *in natural language use*. So, as long as we understand what will constitute a good dose of disenchanting medicine, we can continually explore the frontiers of AI and help ourselves avoid enchantment. More ideas for how to do this will be set out later and, as well as keeping an eye on your spellchecker, you yourself can always be testing Siri and the rest, which are currently hopeless in the face of the tests described in Chapter 10 and, as I argue there, will continue to be so.

To repeat, the reason that the fluency of Ava, Samantha and HAL isn't currently credible is that, in spite of some apparent successes, we don't yet have the ability to embed computers into language-speaking society; it is the language-counterpart of why we have no reason to think that intelligent computers will desire enormous riches and power unless we insert this propensity into them. Brilliant programming can make computers do a lot in the way of simulating language-understanding and responding, as the modern devices show – and they are going to get much better. It is still not quite the real thing, however. But showing that it is not quite the real thing, and it is really important to do so, will become harder and harder and the danger of the Surrender will come ever closer.

Expertise and Writing about AI
Some Reflections on the Project

The trouble with artificial intelligence is everyone thinks they
have something sensible to say about it. On the one hand,
there are those who say that machines could never be creative
or will always lack emotions or consciousness or maybe a
soul; on the other, there are those who insist that computers must be capable of doing anything that we do since we
are machines ourselves. You can hear all this in the pub or
café. At the other end of the spectrum, consider the letter to
the newspapers about the Singularity, written by Hawking
and others. What special expertise do mathematicians and
physicists have to put them in a position to make the kinds
of claims made there? As someone working at the frontiers
of AI said to me recently: 'I wish Hawking wouldn't write
about AI; we don't write about black holes.' So what about
someone like me? In this chapter, among other things, I'll
try to justify what is going on in this book; after all, I don't
actually build intelligent machines so what right do I have to
talk about them?

What do I mean by 'cannot'?

Two central negative claims were set out at the start of this book. It would be understandable if those who have spent their careers writing programs and making machines work better, sometimes in ways that confounded previous generations of outsider sceptics, objected to yet another round of being told by an outsider that their grandest ambitions are misplaced. But, in spite of the recent success, the history of artificial intelligence indicates that quite a few outsiders have made sceptical claims that turned out to be justified even though they were scorned at the time. Disagreement in science is healthy. The advance of science depends on what the sociologist Robert Merton called 'organized scepticism' and the best science is often done as a result of triumph over the supposedly impossible; if this is going to happen, someone has to say what's impossible.

That said, we know that anti-technology prophecies ('humans cannot travel faster than 30 miles per hour', 'heavier-than-air flight will never be achieved by humans', 'no human will ever leave the Earth's atmosphere or gravitational field') are often confounded by technological breakthroughs. Sometimes critics are confronted with Arthur C. Clarke's so-called 'First Law':

> When a distinguished but elderly scientist states that something is possible, he is almost certainly right. When he states that something is impossible, he is very probably wrong.

But physical scientists, who tend to be quite fond of Clarke's First Law, are not known for their backwardness in stating impossibility principles of their own. There is, it has to be recognized, an unspoken disciplinary snobbishness when it comes to saying what can and cannot be done: so far as I

Table 2.1 Some types and examples of impossibility claim

Types of cannot	Example/s
Logical impossibility	We cannot have our cake and eat it too.
Scientific principle	We cannot travel faster than light.
	Perpetual motion is impossible because of the second law of thermodynamics.
	Quantum theory is flawed because the Einstein, Podolsky, Rosen (EPR) paradox shows it would lead to non-local entanglement.
	Paranormal forces are impossible.
Logistic principle	We cannot enumerate and store all possible chess games.
Logistic practice	We cannot build a tunnel between England and Australia.
Technological impossibility	We cannot converse with anyone more than a mile away.
	We cannot make rechargeable car batteries with the energy density of a tank of hydrocarbon fuel.
Technical competence	We cannot translate the Rosetta Stone.
	We cannot make room-temperature superconductors.

know, no one quotes Clarke's First Law when someone says it is impossible to travel faster than light or build a perpetual motion machine.

In Chapter 1 I talked about my negative claims being similar to claims about death and chromosomes. Table 2.1 presents and classifies more types of impossibility claim from strongest at the top to weakest at the bottom. The table deliberately includes some claims that have proved incorrect.[19]

The first kind of claim is generally uncontentious barring some esoteric philosophical treatments. Scientific principles of the kind exemplified in the second line are a favourite of physical scientists in spite of Clarke's First Law. The EPR paradox, which reinforced Einstein in his well-known view

EXPERTISE AND WRITING ABOUT AI

that 'God does not play dice', turned out not to be false, making the resulting discovery of non-locality still more astonishing. The poor old parapsychologists, of course, are always getting it in the neck from outsiders, not least physicists, who, shamefully, are even ready to praise stage magicians for proving parapsychology wrong. Logistic principle seems to be acceptable and so do lots of things you can dream up which violate conceivable logistic practice. The technological impossibility or technological competence claims are more contentious and two of the ones I have chosen have turned out to be wrong: telephones (or smoke signals) confound the distant conversation claim and the Rosetta Stone has been translated. It is not entirely clear what is going to happen to batteries and super-conductors but, if anyone has a theory about those technological achievements being impossible, the dismissive application of Clarke's First Law would be quite out of place.

How about my two sceptical claims? It seems to me that the first of these – 'No computer will be fluent in a natural language, pass a severe Turing Test and have full human-like intelligence unless it is fully embedded in normal human society' – is simply a scientific principle like those in the second row of Table 2.1. In this case, however, being about a technical matter and coming from a social scientist, it might make some people uncomfortable and more inclined to quote Clarke's First Law. Incidentally, I think most protagonists of deep learning (Geoffrey Hinton aside, see Chapter 6) accept the first claim.

The second claim – 'no computer will be fully embedded in human society as a result of incremental progress based on current techniques' – is much less secure. Nevertheless, it belongs among the kind of claims found in the bottom two lines of Table 2.1 and will be supported in the chapters that follow. I could be wrong, but a negative claim like this makes the future more exciting rather than less. Let us, then, introduce Collins's First Law:

> Arguing that impossibility claims should not be made because they have turned out wrong in the past substitutes academic authority for thought and analysis.

Incidentally, let me make clear that neither of my claims is a 'prophecy'. A prophecy has to deal with the *future*, but my claims, along with most of the impossibility claims in Table 2.1, are not prophecies because they deal with the *foreseeable future*. An impossibility claim is about whether one can get to some point by incremental change given what we know now, and I will say we can't, whereas deep-learning enthusiasts will say we can. But my first claim could be confounded by some new and unforeseeable principle to do with how human knowledge works. My second claim could be simply wrong: that would be very interesting. I cannot foresee that happening or I would not have written the book.[20] One has to be careful about the status of an impossibility claim.

Expertises and academics

Currently, there is a competition between AI enthusiasts on the one side – supported by philosophers of evolution and the like, who believe humans are no more than machines – and the critics on the other. That is potentially a good thing but the interplay of creation and criticism has to be done in a certain way if its productive potential is to be realized; too often, it is done badly. The debate is unproductive when the parties direct their arguments primarily at some wider audience rather than at the scientific opposition. A productive scientific debate has to be primarily inward looking

– engaging with opponents in the scientific community rather than sidestepping the hard arguments by aiming outward at the general public; convincing scientific opponents is, as it happens almost impossible, but trying to convince them is the way to sharpen the arguments and enhance the understanding of both sides. Managing such a productive debate can be hard. For example, physics research is often on such a large scale, so expensive, and so esoteric that physicists cannot find anyone to provide serious opposition outside of their own specialist groups. So physicists set up opposed parties within their own teams. In gravitational-wave physics, a field I have studied for forty-five years and will refer to frequently throughout this book, it was always the insiders that were their own harshest critics, casting aside one claimed discovery after another for fifty years until they considered they were ready to agree that they had found the real thing. Social relations in the AI business should go the same way; the creators of AI need to be at the forefront of the criticism of AI, too, because only they have the knowledge and understanding to make criticism as ruthless as possible. If true general intelligence is to be found, every imposter's flaws will have to be exposed in the same way and it is going to be the proponents and producers of the products that will have to do much of the work. Getting people to accept that general intelligence has been produced should be like getting people to accept that gravitational waves have been discovered: not a competition to see who can invent a trick that will convince people, but a painful, even masochistic, exploration of every conceivable doubt.

Artificial intelligence belief

Criticizing AI is not the same as the self-affirming practice of worrying about the dangers of AI. The much publicized concern over the Singularity actually affirms the power of the product in the very process of pointing to its dangers by making us take computers too seriously. It is much more difficult to look inward at the weaknesses of the product and far less fun for the enthusiasts.[21] Yet what we understand to be the actuality of AI is an integral part of its impact! Because of the power of words and public opinion when it comes to AI, part and parcel of reducing the impact is advertising the failures of the product alongside the market-driven urge to advertise its successes and the ego- or ideology-driven urge to triumph over the doubters.

Maintaining disenchantment is going to be especially hard in the face of what we'll call 'Artificial Intelligence belief'. AI belief has a quasi-religious (or counter-religious), ideological element which turns on seeing humans *as* machines, not just something to be mimicked by machines. Its origins lie outside artificial intelligence research, though insiders like Marvin Minsky, with his early 'meat machine' slogan, did not help. The AI community has experienced wave after wave of 'hype' when one kind of innovation or another has promised to solve the problem of human intelligence: my 1990 book is a response to the overblown claims that came with the expert-system revolution in AI. But the community has now experienced many dispiriting flops, expert systems among them. Many AI scientists fear that yet another 'AI winter' will follow the latest deep-learning bubble. The large majority of AI scientists want to get on with building devices that work better, and will help humans run their day-to-day lives better, rather than take over the world or prove that humans are merely machines. Indeed, the strongest AI *belief* seems to come more from philosophers, evolutionary biologists or other outsiders, suffering from the web of enchantment that distance from the frontiers of the technology can weave, and sure that humans can be no more than organic machines designed by the 'blind watchmaker'. Under this ideology, humans are essentially individualistic and competitive, the theory of evolution fitting the theory of free-market capitalism; we machines were born out of evolutionary competition and it is natural for us to continue to follow the competitive rules of the market.[22] The argument becomes political and, intentionally or not, it lines up with individualistic Politics with a large 'P'. We have to steer our way through this mixture of faith, politics, Politics and technology. My characterization of AI belief may be rough and ready; it is partly a product of AI's high media profile and the semi-popular books written by the enthusiasts. But, once more, it is 'the use of words and general educated opinion' with which we are concerned and the public face of AI has to be a feature of the argument.

When AI was an orphan subject there was not much in the way of self-criticism going on because the wagons were circled in defence against a hostile outside world. But nowadays the proponents of AI are led and supported by capitalist emperors. The market has made those working with Google, Microsoft, Facebook, IBM and the like rich beyond the dreams of kings because of what they can deliver to the market place. This means that, unlike other research scientists, they do not need to answer to the taxpayer for the way they spend their money. Suddenly, the enthusiasts have all the resources they need to pursue their dreams. This makes the need for properly organized and run scepticism still more pressing, but instead – and in this respect the only fault of the emperors is the sin of omission, in that they've done nothing about it – we have fairground gimmicks in which computers are ranged against each other in attempts to reproduce human conversational ability and, year after year, some hype-merchant will make the nonsensical claim that the feat has been achieved, in spite of there being almost no serious attempts to work out proper experimental protocols, nor to think about what it would really mean to pass a proper test. As we will see in Chapter 10, the situation is at last being remedied due to the first serious Turing Test-type competition, with a reasonable protocol, having taken place in New York in 2016 (but the powerful groups in the field are not involved).[23]

AI expertise

My academic expertise includes the analysis of expertise itself.[24] Maybe a bit of expertise about expertise can be applied to the relationship between insiders and outsiders in the AI debate? Different kinds of knowledge-making activity each have a different 'Locus of Legitimate Interpretation' (LLI). The LLI is the location within society where legitimate criticism is expected to come from. For instance, in the arts – and the same goes for food and wine criticism – the LLI is widely distributed, for example to newspaper critics and the public; while in respect of fine art, one is entitled to say, 'I may not know much about art (food/wine) but I know what I like.' Though that expression is sometimes used as a joke it has a kernel of truth. Try it on the sciences, however: 'I may not know much about black holes but I know what I like'; such a claim is always a joke. In the case of the sciences, legitimate criticism is restricted to other technical experts; the LLI is narrow. Something pathological is going on in, for example, vaccination revolts, when the general public starts to consider that it has valuable expertise in respect of whether a vaccine is dangerous.[25] Around the time of writing (July/August 2017), a baby called Charlie Gard was kept alive in the UK for weeks against all medical advice because of public protest. As one of the involved medical staff said:

The little child . . . had in effect been kept alive for people such as Donald Trump, the Pope and Boris Johnson [a publicity-seeking British politician], who 'suddenly knew more about mitochondrial diseases than our expert consultants'. (*Guardian*, 5 August 2017, p. 1)

An enterprise like architecture has another kind of LLI, with a lobe that reaches into engineering where only a narrow group of experts is entitled to provide an opinion about whether a design will stand up, and a lobe that also reaches into the arts because architects try to design beautiful buildings and the general public certainly has a say as to whether such buildings are beautiful and/or functional. So architecture's LLI is under tension, illustrating how LLIs can span disciplines and constituencies.[26]

Artificial intelligence exhibits yet another pattern. Though everyone feels they have a right to say something about it, they don't. The locus of *legitimate* interpretation does not reach into the public, whatever gets said in the pub or café, and it does not reach as far as any clever mathematician or physicist. Having said that, legitimate expertise in AI, especially when it comes to belief, does cut across disparate groups of experts, some of whom do not talk much to each other. This is because there are some who have little or no technical skills in the *creation* of intelligent machines, but who do have expertise relevant to the question of whether AI's goals are being achieved. The reason is that AI is engaged in the business of trying to replace, or mimic or reproduce another activity, and the other activity is not the special province of those trying to *do* the replacing, the mimicking or the reproduction. In the case of AI, anyone who understands the relationships between the body, society and knowledge does have something to contribute because they know what it is that has to be replaced, mimicked or reproduced; no amount of computer-related

Table 2.2 The peculiar range of expertises that contribute to the AI belief question

Experts on MACHINES		INTENDED TO MIMIC OR REPRODUCE	Experts on HUMANS	
Designers and Programmers	Siliconeers		Philosophers e.g. the body	Sociologists e.g. The collectivity
Web miners	Software testers		Psychologists e.g. Emotions Consciousness	Neuro-scientists The human brain

technical skills can eliminate the need for expertise in how human knowledge works. To repeat, the reason there is this wide range of applicable expertises in AI, and will be until everyone agrees that computer scientists have succeeded in creating full-blown human minds and therefore they become the undisputed experts in what it is to be human, is that one group of experts is trying to make something that duplicates or mimics something else; that means that the experts on the 'something else' have something to say.

Some of the actual or candidate contributors to the question of artificial intelligence are set out in Table 2.2 – which will be referred to throughout the book. The eight boxes show some of the groups which inhabit AI's locus of legitimate interpretation.

The experts on the left are the program designers and programmers themselves, exemplified by the competing designs of old-fashioned symbol-processing versus neural nets and precedent-based processes; what are referred to as 'siliconeers' are those such as Ray Kurzweil, at the time of writing the Director of Engineering at Google and author of enthusiastic and forward-looking books on AI, who, among other things, are good at understanding the exponential growth of the capacity and speed of computer hardware; there are web explorers, who design programs to exploit the vast quantities of material available on the internet and other distributed sources, Google being the pioneer; and there are software testers, at least some of whom understand the confrontation between our needs and what the pioneers produce. On the right hand are psychologists, who know things about human knowledge and who study emotions and consciousness; philosophers, who also have something to say about consciousness and other aspects of human cognition such as its relationship to the body; neuro-scientists who study how the human brain works; and sociologists, who understand how knowledge develops in societies and how different societies, or sub-societies, with different cultures, interact, and how to learn the content of those cultures.[27]

The *ideology* of AI is not located neatly on the one side or the other. Some philosophers and psychologists from the right-hand side are champions of AI *belief* and some of their understandings of psychology and the brain are used to suggest avenues of research to those on the left. But there is a rivalry and argument between a few of the very ambitious people on the left-hand side and a few of those on the right-hand side. This book is part of that argument but, I hope, a constructive part.

Sometimes the tendency of some AI enthusiasts to extrapolate too far from what has been achieved stands in the way of the easing of relations between left and right. It is easy to read too much into the way the right-hand experts have turned out to be wrong from time to time: it is said that because the right-hand experts have made some incorrect claims that have been confounded by left-hand success, the left-hand experts' more ambitious claims must be accepted; this is rarely the true consequence of these mistakes. Success at chess is an iconic example; it will be discussed later in the book. But, as suggested above, there will be an increasing need for the left-hand experts to cross over and help support rather than attack some of the arguments that typically come from the right-hand side, if the Surrender is to be avoided.[28]

Since I think I am an expert on expertise, I always have to consider my own expertise very carefully when I write about any field that is not my own. I probably know as much as anybody about what it is to do a case study that turns on technical understanding of a field that is not my own as a result of my forty-five-year study of gravitational-wave physics. I have twice passed an Imitation Game as a gravitational-wave physicist (see Chapter 4). My books on gravitational-wave physics are full of simple exposition of the technicalities of the physics, carefully checked by the physicists. In comparison, as Table 2.2 suggests, my technical understanding of AI is poor; I could never pass as an AI specialist in an Imitation Game. Here's the paradox: I almost never offer an opinion on the physics of gravitational-wave detection, leaving that to the physicists, yet in this book I am telling AI specialists about the limitations of AI![29] That's strange. The resolution, of course, is that this book turns on the nature of human knowledge, not the details of AI programming. That said, I know enough about deep learning to be able to say that there are currently no other AI approaches that hold the promise of deep learning, nor approach deep learning in its depth of interaction with society, given the way it draws from the internet.[30]

The future and points of principle

A final caveat is that the main focus of this book involves points of principle rather than today's actualities; the argument looks towards a time when the current limits on the capacity of computers and of human programming skill are even less of an obstacle than they are now. The argument starts from the view of how things might be if everything technical that we can foresee were to come to pass. So, for example, I will, at various points, imagine a world in which not only are computers capable of storing and transcribing everything that is transmitted in written form over the internet and everything that is said by one human to another within the 'earshot' of a telephone, or, given lipreading technology, in sight of a camera, and, of course, everything that is broadcast on radio and television; we are going to imagine, in other words, that nearly all human conversation, from the distant past all the way up to right now, will become available to machines that can recognize patterns and that can be used to indicate what might be being said by one human being to another. It is possible, then, that sometimes the positions discussed here will seem credulous because we are going to assume that certain current technical problems have been solved even if, currently, we are not so far forward as some enthusiasts claim. The points of principle are being raised here in the service of an inward-looking argument – to show what AI cannot achieve, not what it can. Or, at least, they are raised in the service of showing that we have used as much vigilance as we know how in estimating whether the loudest and most optimistic claims are justified.

3

Language and 'Repair'

There is a weakness in Chapter 1 in my use of the artifictional intelligence films and their meaning. The readiness and constant preparedness of humans to be charitable when faced with less than perfect conversational performances and to hear them, nevertheless, as perfect shows us that we could be satisfied that even imperfectly fluent computers are as good at conversation as the best of humans. Normal human conversation is full of breakages, mumblings and every other kind of imperfection, and if we did not constantly 'repair' these faults without noticing there would be no fluent conversation. Therefore, in the normal way of things we take even badly broken conversation as signifying fluency and humanness. We say that Ava and Samantha and Hal are fully fluent but we haven't seen their fluency tested to breaking point. That is the danger we face with the increasing capability of Siri and the rest, and that is why we need to become more and more vigilant; partial fluency too easily passes as full humanity. So the fact that we tend to judge humanity by its linguistic fluency is a terrible danger as computers become more fluent. I am going to back this up with a quotation referring to myself and co-author, but taken from a paper by a computer scientist, so we can see that the problem is already beginning to be recognized.

> This fundamental property of interaction with machines is described by Collins and Kusch (1999) as Repair, Attribution and all That (RAT) – human users constantly 'repair' the inadequacy of computer behaviour, then attribute the results to intelligence on the part of the machine, while discounting the actual intelligence that was supplied in the process of repair.[31]

How misspellings and the like are dealt with by humans

To repeat, ordinary human speech is faulty and broken: we mumble, we don't complete words and phrases, and there is background noise and cross-cutting conversation. That this is the case is made evident to users (and developers) of automated speech-transcribers, who find that background noise or a change of accent or intonation will cause the transcriber to make a string of errors. This is because such speech-transcribers, at least the early generations, work by recognizing patterns of *sound*, not speech, and the sound-patterns are confounded by cross-cutting noises. This reinforces the claim that in ordinary speech, which is con-tinually beset by cross-cutting and damaged sounds, we are always 'repairing' what we hear and making plausible sense of what we hear – by trying to understand it. If we only listened to the sounds we would be as bad at speech recognition as the first generation of automated speech transcribers.

One can illustrate the phenomenal power of our ability to

repair broken conversation through reference to its meaning
by trying the same thing on the page, using broken text.

> at aoccdrnig be olny rsceareh a it mtaetr in oerdr the waht
> dseno't to ltteres a are, the iproamtnt pclae Uinervtisy, taht
> tihng lsat is the and wrod Cmabrigde ltteer in the in rghit
> frsit.

You, reader, probably could not make much sense of that.
But if you are a good English speaker you will be able to read
the following, which is made up of the same 'words', without
difficulty, because you can make sense of it.

> aoccdrnig to a rsceareh at Cmabrigde Uinervtisy, it dseno't
> mtaetr in waht oerdr the ltteres in a wrod are, the olny
> iproamtnt tihng is taht the frsit and lsat ltteer be in the rghit
> pclae.

Sense-making followed by repair is how humans handle language.[32] Incidentally, the first passage shows that the claim
made within the quotation is incorrect – for most of us, more
than the preservation of the first and last letter of each word
is needed if the words are to be understood – sense-making
has to come first.

This principle of sense-making followed by repair
applies, as a matter of fact, to much more of our lives
than speech. Indeed, we impute all kinds of sense to things
even when it is not there. This is most obvious in the case
of anthropomorphism, the imputation of human-like sense
and sensibility to animals and even inanimate objects. For
example, certain car owners are emotionally attached to
their vehicles. The same principle is what makes it possible
for all kinds of confidence tricks to work. Confidence tricksters rely mainly, not on their ability to assume the persona

of someone they are not, but on the willingness of the rest of us, or the particular 'mark' in the cross-hairs, to ignore mistakes and make sense of what is being presented. We found in one of our studies that bogus doctors succeed in hospital environments because the nursing staff will correct errors, assuming the bogus doctors are just out of medical school or trained under different regimes, and this gives time for the bogus doctors to learn on the job and eventually become proficient.[33] Human life, then, is filled with continual repairs of broken speech or faulty actions and the attribution of sense where there is none. It is hardly surprising that we are inclined to attribute sense and capability to machines, including speaking machines, where there is none. And that is why the 1960s experiment with ELIZA, a supposed mechanical replacement for a psychoanalyst, which provided keyboard-mediated question-and-answer consultations, resulted in the machine being perceived as a successful therapist by some patients, even though it was a toy.[34]

But now let us illustrate how easy it is for experts like me, drawn from the right-hand side, to be surprised by what can happen on the left. Let us go back to the broken text about research at 'Cmabrigde Uinervtisy' and consider the first version of it, with jumbled word order; it is very difficult for most of us to repair because it cannot be made into sense. It is, however, a trivial task to repair it using simple analysis of the patterns that can be found in the existing corpus of language. The first program I located in my Google search for anagram-finders (<http://www.litscape.com/word_tools/contains_only.php>) unambiguously transliterated every garbled word but two – 'dseno't' and 'Cmabrigde' – which contain an apostrophe and a proper name (this could easily be remedied). The program was, of course, indifferent to whether the words appeared in the first passage which didn't make sense, or in the second which did make sense. Nearly every garbled word gave rise to only one anagram – the correct one – with the one or two remaining ambiguities being resolved by reference to the fact that first letters from the original words are preserved; we did not even have to resort to the last letter condition. So there is enough in the statistics of existing English words to enable a computer to transliterate jumbled words without any reference to sense and without understanding anything. We humans, on the other hand, do need to make sense of the passage to do the repairs, as the differential comprehensibility of the two versions of the passage shows. The understanding of the nature of repair belongs to expertise found on the right-hand side of Table 2.2 (p. 30), and the way it could interact with expertises from the left is illustrated by the different solutions to the garbled word example. This little experiment is both a surprise and a warning: it is a warning because it indicates how easily 'brute strength' algorithms – 'tricks' – can solve problems that we think require deep understanding and this can give the illusion that problem-solving computers have deep understanding. (The deep-learning approach has, of course, much greater power than anagram solving.)

The centrality of language

It was Alan Turing who put language at the centre of artificial intelligence but it would have got there anyway. Language makes humans special – it is what enables them to share their ideas, makes the interweaving of their individuality far greater than the sum of its parts, and enables them to want to spend their riches on fancy cars, remodelled bathrooms and taking over the world, or to prefer poverty and contemplation; it also makes it possible to engage in telephone sex. In other words, language is the medium of culture, and culture is what makes us more than an elaborate set of potentials or just complicated animals.[35]

To understand what artificial intelligence is and where it is going it is first necessary to understand the relationship between computers and language. It is necessary, among other things, to see that the step from a computer that cannot deal with something like 'I am going to misspell wierd to prove a point', to one that can deal with this and similar problems, is a huge one; it's not just a matter of improved spellcheckers. To repeat the point made earlier, repairing that mistake in particular is easy once it's been spotted and the 'fix' programmed in; the problem is to correct all such mistakes within the ever changing flux of language as it is spoken and written in society, and that requires being embedded in the language-speaking society. Though computers get close to mimicking some of these abilities, they are still well short.

One of the surprises with the advent of precedent-based methods and the growth in computer capacity and power has been a big advance in language-handling ability. The capabilities of the latest voice-interactive devices go far beyond anything I could have foreseen a decade or two back. At the time of writing this passage (November 2016) another advance has been claimed. It is said that computers are now just as good at *transcribing* speech as human transcribers. If that is true (and I doubt it), it will be marvellous given that I have to do a lot of laborious transcription in my work – for example, I must transcribe tape-recordings of various different speakers in noisy environments; existing transcription programs cannot cope with these. Could it be that this latest generation of machines really has cracked the problem? I read that the latest programs make mistakes, but no more mistakes than human transcribers. If that is true, artificial-intelligence science and engineering will have achieved something

remarkable and something I can use. But before we know
whether to count the achievement as satisfying the claims of
mimicking human-like intelligence, we'll need to know whether
the remaining mistakes made by the machines and the few
mistakes made by humans are of the same kind. If humans
and machines make systematically different kinds of mistakes
then this engineering marvel, wonderful though it would be,
brings us no closer to human intelligence than did the, at the
time equally wonderful, pocket calculator.[36] Unfortunately,
when, in August, 2017, I tested Google's transcribing device
while sitting with one of the founders of deep learning (we
used his smartphone), we discovered that it did make mistakes
in a noisy environment – the kind of environment where
humans would not make mistakes – so it looks as if my doubts
are well-founded and devices that can do my tricky transcribing are not here yet. This is a good test for computers and
we'll mention it again at the end of the book. Though current
transcribers are a *huge* improvement over the previous generation, and a wonderful technical achievement, this book is
demonstrating that if you want to think properly about what
has been achieved in relation to human abilities you have to
see the glass, not as half-full, but half-empty.

Getting language into computers and the Chinese Room

How do you teach a computer to use language? You can
program it to construct grammatical sentences, you can teach
it spelling and the dictionary definition of words or you can
get it, like ELIZA, to recast the sentences that humans type
into it, but all this is much more difficult than it seems. ELIZA
tended to make a mistake after about two exchanges but, to
repeat, the deeper point is that in natural language use the

rules can always be broken and new precedents can be set. Our simple sentence breaks the rule for the correct spelling of 'weird' but it is, nevertheless, a sentence that is fully comprehensible to humans. The point is made still more strongly by the garbled spelling of the sentence about research on spelling at 'Cmabrigde Uinervtisy', though the spelling is exactly as it is intended to appear in this book. And here, to show the open-endedness of the point once more, is another reasonable sentence that shows how to break grammatical rules but can still be understood: This time, at the end of the phrase, the verb I will put, German word order to demonstrate.

The Chinese Room revisited

Because we don't want to get into programming technicalities we'll use a famous thought experiment as a way into thinking about language and computers. Using this thought experiment, which was developed in 1980, long before deep learning came along, will also reveal how times have changed. We'll show in later chapters that one out of the three problems that, according to me, beset this thought experiment, has been overcome or ameliorated with the automated analysis of huge databases in near real time.

One of the most long-running philosophical debates inspired by the Turing Test is John Searle's 'Chinese Room' argument; it is not on many people's minds these days but it can still do a useful job for us. It was intended to turn analysts' attention to the question of whether machines are or can be conscious; the question is all about whether executing code to produce an answer to a question is different from answering that question through conscious understanding. There was an enormous debate about this. Here, we are going to use the idea of the Chinese Room for a different purpose – to bring out some questions about language.[37]

Figure 3.1 *The Chinese Rooms*

At the centre of Searle's thought experiment is a room with a huge 'look-up table', with inputs on the left corresponding to responses on the right. On the left side of each entry in the Chinese Room's look-up table is a question written in Chinese and on the right side is an appropriate answer, also written in Chinese. Chinese speakers approach the room and post questions written in Chinese through a letterbox. Inside the room is an operator who does not speak Chinese – Figure 3.1A. The operator matches the posted questions to one of the shapes of the squiggles on the left of the look-up table, without understanding them. The operator then carefully copies the corresponding squiggles from the right side – the squiggles that represent an answer to the question – on to a piece of paper, and passes it out through the letterbox. To the person on the outside the question has been answered; what has happened is analogous to running a piece of computer code. We may imagine – and here we are elaborating on Searle's design to make it correspond more with the Turing Test – that side by side with the Chinese Room is another room with a letterbox, inside which there are no look-up tables but there is a wise Chinese speaker (we represent wisdom with a capacious beard); let us call it the 'control' room – Figure 3.1B.

If the same questions are posted into the letterbox of the control room, the Chinese speaker will respond by writing out equally appropriate answers; this time, however, the Chinese speaker will have answered using their understanding of Chinese. The premise is that questioners will not know which room is answering via look-up tables and which via the conscious understanding of the wise Chinese speaker, and that premise is fine as far as the aim of the original thought experiment is concerned. The premise is equivalent to saying that, for the purposes of argument, the Chinese Room will pass the Turing Test. What Searle was proving was that a competent linguistic performance, as in the case of the Chinese Room or Turing Test, does not show that a computer is conscious or that it is operating in the same way as a human.

Searle's argument would work just as well if the Chinese Room was not full of look-up tables but contained only one strip of paper with the answer to a single question written in Chinese: let us say the answer is 'light takes eight minutes in its journey from the Sun' (but it could be anything). Now the answer-seekers are told that the Room can answer astrophysical questions in Chinese and that they can test it by passing in a strip of paper with the question 'how long does light take to get from the Sun?' They push the paper through the letterbox; inside is a pigeon trained to pick up the only other piece of paper in the room when it sees a piece come through the letterbox, and push it out; on that piece of paper is written the answer to the question. In the control room, 3.1B, is the wise Chinese speaker, and whether the questioners ask the 'Pigeon Room' or the control they get the same answer. The Pigeon Room poses exactly the same puzzle about the difference between a mechanical performance and a conscious performance as the original Chinese Room – there is no need for all those exhaustive and exhausting look-up tables to make the point, which is, indeed, intended to apply to all computational processes. Unfortunately, the look-up tables have the potential to confuse people into thinking that a look-up table-based language-responder, if it was big enough, would actually be linguistically fluent.[38]

Searle probably wanted to make the relationship with the Turing Test obvious, so he felt he had to set up a mechanism that would perform as well as a genuine Chinese speaker in a Turing Test – it had to be able to answer any question. That, perhaps, is why he did not make the point with something like the pigeon and a single question and answer. But, rhetoric aside, the complexity is irrelevant as far as Searle's main point is concerned.

The more interesting question from the point of view of this book is whether a Chinese Room based on look-up tables would pass a maximally demanding Turing-type test – we'll start to introduce the Turing Test a few headings below. It is important to know the answer to this question because if it could pass then we would have a possible route towards AI. There are, however, at least three arguments that suggest that it could not pass. These arguments are useful for understanding the extent to which deep learning has gone beyond the kind of AI that was under discussion in the high days of the Chinese Room argument in the 1980s.

First Chinese Room problem: Proliferation of misspellings

Nowadays we have a new way to imagine how the Chinese Room's look-up tables could be constructed. Let us imagine that we scan the entire existing internet and the corpus of all electronically mediated conversations for Chinese questions with answers and write them all out in the look-up table. It will be pretty big but it will be small in terms of the universe. But now let us allow that when the questioners write in their questions they occasionally make spelling mistakes. Indeed, let us suppose that spelling mistakes and the like are deliberately inserted so as to give the 'computer' the opportunity to demonstrate its human-like capacity for repair. The wise Chinese speaker in the control room will be able to cope with that problem easily – he or she will 'repair' the misspellings and produce an appropriate reply (which might also contain the occasional misspelling since he or she is human). But the Chinese Room look-up table, even if it is constructed by stripping all question-and-answer turns from the existing recorded corpus of Chinese, is most unlikely to find anything on the left-hand side that corresponds to newly misspelled or otherwise damaged questions: there is most unlikely to be anything that exactly matches the misspelled squiggles in the question because the number of ways of creating misspelled questions is open-ended. If the look-up table version is to work, the left-hand entries would have to include not only all possible questions but all *possible* mangled versions of those questions, not just all mangled versions of those questions that have already been uttered or written. For each correctly spelled question there are an indefinite number of possible mangled versions, limited only by the range of human error and ingenuity. For example, the left-hand side needs to include the Chinese equivalent of 'I am going to deliberately misspell <u>wierd</u> to see if you can read it'. It will also need to include the Chinese equivalent of 'This, the order of words in German sentences illustrates'. And it will have to include the Chinese equivalent of:

> Can you read this? aoccdrnig to a rsceareh at Cmabrigde Uinervtisy, it dseno't mtaetr in waht oerdr the ltteres in a wrod are, the olny iproamtnt tihng is taht the frsit and lsat ltteer be in the rghit pclae.

And this:

> Can you read this? at aoccdrnig be olny rsceareh a it mtaetr
> in oerdr the waht dseno't to ltteres a are, the iproamtnt pclae
> Uinervtisy, taht tihng lsat is the and wrod Cmabrigde ltteer
> in the in rghit frsit.

And it will have to include the Chinese equivalents of all the possible versions of those sentences that human ingenuity might come up with, and all the equivalents that correspond to every other mangled passage that the inquirers might care to offer up.

So the left side of the tables will expand enormously if the Chinese Room depends on exact matches with the examples passed in – there will almost certainly need to be more entries than there are particles in the universe.[39] Thus, the Chinese Room simply will not work – fixed look-up tables cannot substitute for real-time sense-making. Once more, we are getting at the open-endedness of creativity in language, and humans' ability to make sense of it nevertheless.

Second Chinese Room problem: Archaism

The second argument is easiest to make by thinking in terms of two differently configured 'control rooms'. Remember, the control room contains a wise Chinese speaker who answers the questions using his or her conscious native understanding. Now let us imagine we have two control rooms with two wise Chinese persons who we will call 1B and 2B, both recruited as adults (see Figure 3.1). The difference is that 2B is never allowed out; the room is fully equipped with a bathroom and kitchen and food is passed in and waste passed out, but there are no newspapers, television or other media, and no one ever again speaks to 2B; 2B is isolated from Chinese society from the moment he or she is shut into the prison-like control room. Wise Chinese person, 1B, however, has a TV, goes home every night after work and remains fully embedded in Chinese society.

Over time, the performances of 1B and 2B will diverge because the Chinese language will be changing but 2B will not keep up with the changes. So the performance of the control room containing 2B will slowly become archaic while the performance of 1B will remain current.

At best, the Chinese Room, with its fixed store of look-up tables, can perform no better than the control room containing 2B and, over time, will start to fail Turing Tests conducted with 1B as the comparator. It is clear, however, that the right comparison in a Turing Test has to be with 1B because 2B is not a well-socialized human being: if humans are to continue to understand ever-changing social contexts they cannot isolate themselves from the rest of society. If we want the Chinese Room to pass comparison with 1B, then someone who *is* embedded in Chinese society will have to keep updating the look-up tables. But if someone keeps updating the look-up tables, it is that person rather than the Chinese Room itself that is supplying what is needed – namely, social intelligence: the Chinese Room in its initial, non-updating incarnation hasn't captured Chinese language, but just a frozen moment of Chinese language. If there is updating going on, however, the Chinese Room is simply the mediator between the person updating the look-up tables and the inquirers: it is just a hugely complicated teletype machine! It does not demonstrate intelligence at all.

Third Chinese Room problem: The vernacular

The third problem for the Chinese Room is vernacular speech – notably, bad language. Presumably, if a machine is to pass as human, it will need to be capable of swearing and cursing from time to time. It would certainly have to cope with inputs that included swearing and cursing, but if it never swore and cursed itself it could not be said to mimic the full range of human conversation. So at least some Chinese Rooms would need bad language on their output side as well as their input side, and they would have to produce it spontaneously from time to time – and quite frequently, if it was to represent certain kinds of human speaker. Without that capacity, the room would not pass a demanding Turing Test. Now, when a simple look-up table picks a response, it will pick it at random, so the likelihood of it returning bad language is equal to the frequency of bad-language turns in the set of responses. As can be seen, there is a good chance that it will use bad language at inappropriate moments.

It may well be that this is, in principle, the problem that was responsible for the failure to cope with vernacular that was demonstrated by Watson, the famous IBM device that won the game of *Jeopardy!* (to which we will return later). The designers of Watson decided to program the *Urban Dictionary* of slang and so forth into Watson, to make its speech more natural. But the new terms included offensive language and Watson, having no understanding of social context, used these with far greater frequency than appropriate, and in the wrong contexts. In *The Atlantic* of 10 January 2013 we read:

> Watson couldn't distinguish between polite language and profanity – which the Urban Dictionary is full of. Watson picked up some bad habits from reading Wikipedia as well. In tests it even used the word 'bullshit' in an answer to a researcher's query. Ultimately, Brown's 35-person team developed a filter to keep Watson from swearing and scraped the Urban Dictionary from its memory.

Once more, the thing that is missing from both the Chinese Room and Watson is the ability to understand the context in which it is working.[40]

Have the problems been solved?

What we have found out by looking at the Chinese Room – and remember, we are not dealing here with Searle's question but our question – reveals some of the problems that will have to be solved if computers are ever to become truly fluent language users and pass the most demanding Turing Tests. They will have to be able to repair non-standard usage and produce it themselves, they will have to keep up to date with contemporary society and they will have to understand context well enough to use the vernacular and the like in appropriate ways, according to the social setting. Some of this might be partly addressable by adding extra rules, but these extra rules won't cope with the developing flux of language and social convention without the computer being embedded into that changing flux. As always, we come back to the problem of embedding in language-speaking societies.

Before going on, let us note that recent developments have partly confounded what I took, some decades back, to be a problem without a solution – the updating of the look-up tables. Because, nowadays, we can plug computers into the internet we have an automated means of continually updating them in respect of the changing way a natural language is spoken in society. We can have something like the Chinese Room (it isn't actually the Chinese Room if it is plugged into the internet), that keeps up with the moving frontier of the flux of conversation. As we have seen, this still doesn't solve the problem of the immediate context – the problem of the vernacular shows that – but it is an interesting development and does solve a little bit of the problem of archaism in a way that could not be foreseen a few decades back.

The Turing Test and its complexities

We are going to want a test for whether computers have solved these problems. In 1950 Alan Turing published a paper entitled 'Computing Machinery and Intelligence'. In it he proposed that we replace vague questions about whether computers are intelligent with a simple test – that is, a test that seems simple until one gets into the detail. We'll introduce the Turing Test here and begin the task of analysing it but return, in Chapter 10, to further elaborations, showing just how hard it is to pass when the protocol is sufficiently demanding.

Turing's test was based on a parlour game in which men and women pretended to be each other: a judge passed the written question to a hidden man and a hidden woman; the woman answered naturally (say) while the man pretended to be a woman. The judge's job was to determine who was who from the written answers. Turing proposed that the man be replaced by a computer, and that if the judge could not tell the computer from the real man after five minutes of questioning we should call the computer intelligent.

The Turing Test seems simple but it's not. First, note that the task of the computer is to mimic, not a man, but a man who is pretending to be a woman, which is much easier since a man, in this role, can be expected to make mistakes. So, to sharpen up the test, we will want to have the computer mimic a person who is being themselves, not someone who is pretending to be someone else. Second, five minutes of questioning is not very long – surely we should have a much longer test than that and surely we should be able to return and try the test on repeated occasions with new dialogues. We know from experiments running our own Imitation Games, which are like the original parlour game, that it can take an hour to run through half a dozen questions and answers, so half a day's test might be a minimum requirement for even one computer on one occasion.[41] It also seems reasonable to expect the computer not just to answer questions, but to engage in human-like conversation, perhaps spontaneously, taking the discourse in novel and interesting directions.

Then we need to think about the other parties to the conversations. Let us start with the judge. Does the judge know that one of the participants is faking? In the 1960s, the famous but very simple program ELIZA was thought to be conversationally competent by some people, and this is the kind of thing that causes certain analysts to say the Turing Test is too simple, but its psychiatric patient-interrogators had no idea they were talking to a computer and so had no idea what they were looking for. That is vital – the judge should know that one of the parties is a computer and that a test is taking place. If the judge does not know that, then unnoticed repairs will be happening continually and passing the test will be trivially easy. A Turing Test is worthless unless the judges are exercising the most determined scepticism and vigilance because, as we have seen, humans continually slip into repair mode without noticing. A still better way of conducting a Turing Test, and I'll argue this at length in Chapter 10, is for the judge to work out whether the computer can repair *his or her* broken input because it is repairing broken input that is the quintessentially human activity. In other words, the judge/interrogator should mumble and misspell – including misspellings that are not meant to be corrected, like the wierd example – and see if the computer can cope in the way that humans regularly cope; that is one of the ways in which Turing Test judges need to be trained. Given the growing success of programs like Siri, which are precedent-based, the questions will need to be carefully worked out.

Notice also that in the 1960s ELIZA case, there was no comparison, just psychiatric patients talking to a hidden machine – but we always need to compare the machine with a human conversational partner. Then, given that the judge *does* know that one of the parties is faking, there is the question of whether the judge is the same type of person that the computer is pretending to be – and, of course, they should be the same type of person. Just imagine a situation where the two conversational partners other than the judge – computer and human – were both monoglot Chinese speakers, then, unless the judge is a native Chinese speaker too, the test is worthless. The same applies to every other human characteristic that the computer is trying to mimic: the judge had better be a quintessential version of that kind of person being tested or the test won't tell us anything. The other conversational partner who is answering naturally – the 'non-pretender' – had better be an almost perfect example of that type of human too.

Because of all these complications, commentators have offered wildly different accounts of whether the Turing Test has been passed and how difficult it is to pass.[42] The view taken in this book is that a well-designed Turing Test has never been passed and that it is not going to be passed in the foreseeable future, and that this should already have been made clear to the public by organized sceptics coming out of the heartlands of AI.

Actually, we are closer here to the view of someone like Ray Kurzweil than might have been expected. Kurzweil does not believe the Turing Test will be passed until 2029. Kurzweil writes: 'There is no set of tricks or algorithms that would allow a machine to pass a properly designed Turing test without actually possessing intelligence at a fully human level.' He is right.[43] Because Kurzweil is such an iconic figure among AI enthusiasts I am going to use his ideas as a foil throughout the book.[44]

This highlights another vital difference between AI pursued as science or engineering and AI pursued as full reproduction of human intelligence. Engineers should be reasonably happy if the programs work on 95 per cent of the tasks set for them or some such figure, so long as we know when and why the programs don't work in the residual 5 per cent. With AI engineering and science, the aim is to make AI useful, not perfect. To make an artificial human brain, on the other hand, reproduction at or near the 100 per cent level has to be achieved. Thus, in a maximally demanding Turing Test it cannot be that a computer is shown to be able to mimic very dull people who have little imagination or fluency; it must be that the apogee of fluency is mimicked – the natural language equivalent of something like beating world champions at Go. More reasonably, the task must be something like being as good at repairing broken speech as the top 0.1 per cent of human editors. While we fall short of this we will not have demonstrated that computers are as good as humans in their cognitive capacities. Unfortunately, some AI believers seem to think that achieving the capabilities of some high percentage of the human race shows that achieving the capabilities of nearly every member of the human race is just round the corner; they object when critics complain about the missing element. But the AI believers have to be deeply concerned about the missing few per cent because their project is binary. There are already lots of things that humans do that machines can do just as well, and often better. Pulling is done better by tractors, travelling fast is done better by cars and trains, calculating is done better by computers, and so on. But it does not follow that humans are simply machines – a combination of tractors, cars, trains and calculators. To support the argument that humans are merely machines would require, at the very least, that machines could do *everything*, or nearly everything, that humans can do in the way of language-handling, not just some of the things. That is why nothing less than 99.9 per cent, (if we are not expecting a Shakespeare play) is good enough for that project. We'll come back to these points in Chapter 10 and look again at just what kind of human performance we should be aiming to reproduce as Turing Tests become more refined.

Nevertheless, because of the centrality of language, that 99.9 per cent performance still refers to language use alone. A test based only on the exchange of typed words will be adequate so long as it is hard enough – we won't need robotics.

The matter is confounded by the complicated way in which language is sometimes believed to reflect physical abilities. Philosophers such as Hubert Dreyfus (see p. 68), hold that fluency in language can be attained only with a human body. I have argued strongly against this, my arguments turning on the idea of 'interactional expertise'. Interactional expertise will be discussed at length in Chapter 4, but it implies that a thorough understanding of language and practical matters can be acquired through immersion in the spoken discourse of practical experts without any need for practice or other bodily engagement. Nevertheless, even I argue in Chapter 8 that enough of a body to allow the kind of social interaction that will *engender trust* seems necessary if contributions are to be made to the kind of specialist knowledge typical of frontier science. If these arguments are true, then a purely linguistic test is automatically a test of bodily capabilities too and, perhaps, this makes a purely linguistic test still more revealing. This is a difficult topic.

Incidentally, to my astonishment, I discovered that my view, and that of many others, on the pre-eminence of language, is not shared by at least some of the very few top pioneers in the deep-learning field – not to mention being incompatible with those of Geoffrey Hinton, probably the most important pioneer of deep-learning neural nets (see Chapter 6). The view held by some deep-learning pioneers is that animal intelligence goes way beyond the intelligence of current computers, but that when computers have achieved the intelligence of, say, orang-outangs – creatures not known for their cultural accomplishments – then the step to the achievement of full linguistic capabilities would be a small one.[45] It seems strange to me that the step to full language from animal intelligence is thought of as a small one when no animals achieve it in spite of domestic animals being fully immersed in human culture and certain apes and the like being deliberately and extensively trained in language with close to zero success. A few minimal symbol manipulations do not amount to success – what about repairs of broken speech, or telephone sex? Remember, the glass has to be seen as half-empty not half-full if nonsense is to be avoided! That animals have language is also at variance with the discussion of the crucial role of language in human brain development in the works mentioned earlier.[46]

In this same conversation the current linguistic abilities of even deep-learning computers was, to my surprise, said to be not a matter of fluency, but of 'tricks'! Claims in respect of language can vary wildly among deep-learning protagonists. Thus, it can be claimed that the most difficult linguistic/-cultural problems I can invent will soon be within the purview of deep learning, while within the same twenty-four hours in the same group, it can be claimed that full language would not be achieved prior to the achievement of animal-like intelligence and the achievement of animal intelligence is still a very long way off. The resolution, I think, is that the early successes of deep learning tend to make the accomplishment of future, much more demanding, goals seem inevitable. We must always remember that many of the claims made for deep learning are not based on current success but a projection from current success, which is a quite different thing as the history of science and technology, not least AI, shows over and over again.

4

Humans, Social Contexts and Bodies

How do humans come to understand context?

The argument is that the central obstacle to the creation of human-like intelligence is the difficulty of creating human-like sensitivity to social context – in other words, a human understanding of society. As already noted, the way humans come to understand social context is through 'socialization' – by being brought up within a society or, in later life, as a result of immersion in some initially unfamiliar community that is to be understood. I can understand what needs to be done with illustrative sentences containing flagged misspellings because I am an expert native English speaker; that is how I have been brought up and that is the society in which I have been immersed for many decades. As a sociologist of scientific knowledge, I also know a lot about how to embed myself and become socialized into communities that I have not encountered before. I also know how scientists and others develop new ways to think creatively, turning what once would have been counted as mistaken views into acceptable innovations: I study how we 'change the order of things'. In the meantime we can reduce the vagueness of the notion of understanding context by defining it in terms of what it does: *understanding context is indicated, among other things, by the ability to distinguish between acceptable rule-breaking and acceptable precedent-setting, on the one hand, and simple errors and mistakes, on the other.* The argument of this book turns on that indicator and, to repeat, embedding in society is the key. To anticipate a point from later in the book (Chapter 8), changing social order often involves fierce arguments among the participating humans, usually in carefully restricted groups, about whether some novelty is acceptable or unacceptable rule-breaking and precedent-setting. A problem for full artificial intelligence will be to work out how to make computers that have the social abilities to be accepted in small, closed, secretive, trust-based groups, so that they can take part in the arguments.

Our criterion for success in creating general human intelligence in computers will be human-like language-handling with context sensitivity – embedding in language-speaking communities. The language-handling criterion is in the tradition of the Turing Test, to which we will return continually. I am arguing here that embedding in societies is a necessary precursor to the successful passing of properly designed Turing Tests.

Working all this out will turn on some subtle distinctions. For example, are we talking of individual computers which embed themselves within, and learn from, existing human communities? Or are we talking of autonomous societies of computers that develop their own languages, with which human societies coexist and from which humans might learn in the ways that we can learn from strange *human* societies with which we come into contact? These differences will be discussed at length in Chapter 5.

The problem of a non-embedded sociology

Now, let us draw once more on a body of experience with human understanding that maps on to the problem of intelligent machines. The problem of embedding in society presented itself long ago to sociology and the social sciences in general. In the 1950s and 1960s, social scientists fought with each other over whether subjects like sociology were 'scientific' or 'interpretative' – 'objective' or 'subjective'. The argument was not very illuminating because no one really knew what the terms meant; the nature of science is much better understood these days and we now know that a 'subjective' method can be perfectly scientific: it depends what the analyst is trying to achieve. But one crucial opposition drawn from the old debate lives on. It is a matter of whether sociology, anthropology, ethnography and related disciplines can be 'objective', in the sense of analytically sound, if the societies being studied are looked at solely from the outside, or whether the analyst must understand the world from the point of view of the 'actors' – the people being studied – before he or she can make sense of a society. This problem has a lot in common with the question of whether pattern extraction and statistical pattern-recognition methods in AI can mimic or reproduce (linguistic) human action in general. If they can, then a purely 'objective' sociology is possible; if they can't, the sociology must begin with subjective understanding.

Let us write some science fiction of our own in the spirit of a thought experiment that bears on this question. Imagine we discovered life on an alien planet – life with, on the face of it, given its level of organization and tool use, roughly the same level of intelligence as ours. But let us suppose the life is based on non-carbon chemistry, the atmosphere of the planet is poisonous to us and the bodily form of the aliens is strange. We find that, try as we might, we cannot understand the aliens' language, nor they ours (more arguments bearing on this point will follow in the next chapter). But they are not hostile and we can hover above them in our spacecraft to observe everything they do, this being easier because they live outdoors in perpetual light – nothing is private.[47]

To try to understand the aliens, we gather huge quantities of data on their movements, including hand and lip movements, and we make endless records of their speech and writing. In other words, we collect, in respect of this alien society, all the data that Google, Facebook, supermarket chains and intelligence agencies would like to capture in respect of our society, except that we cannot, at the outset, understand any of it. The question is, could we make sense of it with sufficiently assiduous study and ingenious statistical analysis?

This, to repeat, is essentially the problem that was thought solvable by those who believed that sociology could be an *objective* discipline, narrowly construed: it was a discipline that could work purely by collecting data about societies from the outside. Thus, without ever engaging in any messy and subjective process of understanding, or 'interpretation', one could eventually work out how an alien (albeit Earth-based) society organized and ran itself. This is also equivalent to the artificial-intelligence-without-social-embedding problem. To a neural net (see Chapter 6), or other such program, human societies are alien societies. But it is believed by some that through collecting enough data about them and analysing them with sufficient ingenuity, though without any guidance from human programmers who understand the societies (that is, through completely unsupervised learning from unlabelled data), the machines will be able to reproduce the patterns in the world which generated the data. Consequently, they will be able to learn to live in that world in a way that is indistinguishable from that of the natives and indistinguishable by the natives from their own way of living – among other things, of course, passing the most demanding Turing Tests.[48]

The position advanced here is that, in contrast to this view, sociology must begin with understanding – with 'subjective' or 'interpretative' methods. The basis of such methods is, once more, immersion in the linguistic discourse and, if possible, the social life of the society being studied, so that 'socialization' takes place. Thus one learns the native language to the point that one understands how native members naturally go on in their day-to-day lives and finds oneself making similar judgements to the native members – exactly the same ambition as we have for intelligent computers. Incidentally, this 'subjective' method, called socialization, is uncompromisingly scientific insofar as anyone else using the same approach on the same materials would come up with the same findings; this means its conclusions are replicable and therefore 'objective' even though its methods are subjective, something that has been the cause of considerable confusion.[49]

It is worth noting that one can explore the sociological problem quite amusingly by investigating human societies with concepts that are mutually alien because they play different sports. So, if we look at the terms 'short leg' (which belongs to cricket) and 'running back' (which belongs to American football), and try some translation and back-translation using Google Translate, we can generate Table 4.1.

Cricket is played in India and South Africa but not in France or China; hence the difference in outcome between the first two rows and the second two rows of the table. American football is not played in India, France or China, but (so I discovered as a result of this exercise) is known in South Africa, and hence the outcomes in the last four rows. One simply cannot say 'at short leg', or 'a running back', in French or Chinese (or 'a running back' in Hindi).

Table 4.1 Back-translations of sports terms using Google Translate (February 2017)

Language	Original	Translation	Back-translation
Hindi	I field at short leg	शॉर्ट लेग पर मैं मैदान	I field at short leg
Africkaans	I field at short leg	Ek gebeid op kort been	I field at short leg
French	I field at short leg	Je plante à la jambe courte	I plant with short leg
Chinese simplified	I field at short leg	我在短腿	I'm on short legs
Hindi	I am a running back	मैं एक बापस चल रहा हूँ	I'm walking back
Afrikaans	I am a running back	Ek is 'n hardloop terug	I'm a running back
French	I am a running back	Je suis coureur	I'm a rider
Chinese simplified	I am a running back	我是跑回	I was running back

The Imitation Game and interactional expertise

In recent years we have developed a way to measure the extent to which a native society has been understood by an outside analyst. The method is the Imitation Game, which is the Turing Test played with humans; once more, the sociological problem and the artificial-intelligence problem reflect each other.[50] In the human-only version, under questioning by a 'judge/interrogator', a person, who is not a full-blown expert in some area of social life, pretends they are an expert; the judge/interrogator is a full-blown expert and so is the person to whom they are being compared and who answers the same questions. The game can be played in many ways – with single individuals or with smaller or larger groups. Thus, we have shown that the blind are much better at passing as sighted than the sighted are at passing as blind, which is exactly what you would expect given that the blind spend much of their time immersed in the spoken discourse of the sighted, whereas the sighted spend almost no time immersed in the discourse of the blind: that was a small-group version of the game. As for the individual version of the game, I have spent forty-five years immersed in the field of gravitational-wave physics and, though not a physicist, have 'passed' an elementary Imitation Game test on two occasions separated by about ten years. On these occasions it has proved difficult to distinguish me from a gravitational-wave physicist when my answers to technical questions in the field, set by a gravitational-wave physicist, are compared to those of 'native' gravitational-wave physicists.[51]

What these tests were originally designed to show is that the language of a specialist group can be thoroughly learned through long and deep immersion in the spoken discourse of that group without needing to engage in the practices of the group. What is learned in this way is 'interactional expertise', a concept which, as time passes, seems to have more and more reach and application. Interactional expertise enables a person to understand worlds of practice within which they cannot actually practise themselves purely by being immersed in the linguistic discourse of an expert group. Our research shows that a person who is fluent in the spoken discourse of a practically skilful community – a person who has acquired the 'practice language' – will tend to make the same kind of practical judgements as the skilled practitioners themselves; this is what the experiments on gravitational-wave expertise revealed. These results indicated that fluent language use is itself a skill which carries with it various tacit capacities, those related to technical judgement being similar to those acquired through physical practice itself; this shows, inter alia, why the Turing Test is potentially very powerful; it can test practical understanding through language alone.[52]

Thinking further, it becomes clear that without interactional expertise society would cease to function because no practical specialist would understand another practical specialist and they would never be able to coordinate their activities. Thus, to return to the blind, a blind person can fully understand tennis so long as they have been immersed in the spoken discourse of tennis, even though they cannot play it. That's also why 'Samantha's' virtuosity when it comes to telephone sex is imaginable (assuming that the problem of fluency had been solved), in spite of the fact that she doesn't have, and never has had, a body; the same goes for 'Ava' and 'HAL'; the body isn't important, as it is immersion in the spoken discourse that is important. In sum, when it comes to an individual learning and understanding the language pertaining to a world of practice, computers do not have to practise themselves – what is important, as we might say, is the mind in the machine so long as it can speak. The concept of interactional expertise explains why the focus throughout this book is on language-handling. Language-handling is the quintessential feature of human intelligence that enables both society to work and humans to be what they are (cf. playing board games). Under this model, contrary to what has been claimed by some deep-learning experts, the creation of robot animals, while it may not be easy, is easier than mastery of fluent language in even non-robotic computers because the former does not require embedding in the immensely complex social life of humans, whereas the latter does.[53]

Mimeomorphic and polimorphic actions

To repeat, the central claims of this book are that computers will not be able to behave like humans and that their intelligent performances will not be indistinguishable from those of humans unless they are embedded in language-speaking societies. I have claimed that the same principle applies to the social sciences – the typical actions of humans, leave alone their innovatory actions, cannot be understood without immersion in the corresponding societies; they cannot be understood from the outside by observation and statistical analysis. This is such a crucial claim that it is worth explaining it in yet another way, drawing on the distinction between 'polimorphic' versus 'mimeomorphic' actions.[54]

An action is anything that a human does *intentionally* – as opposed to accidentally or caused by reflex or outside force. Falling over is not an action but diving to the ground is an action; blinking is not an action, winking is an action; mistyping 'weird' as '<u>wierd</u>' is generally not an action but an unintended mistake, although, as encountered in this book, it is an action.

Mimeomorphic actions are carried out by executing the same externally visible behaviour every time. In the normal way, spelling 'weird' is executed by causing my fingers to go through the same behavioural routine every time: *weird, weird, weird* ... you can visualize my fingers doing the same dance on the keyboard on each repeated occasion. Other examples of mimeomorphic actions are saluting, or similar parade-ground manoeuvres, synchronized swimming, dressage, dialling a telephone number and so on. Note that we do not have to succeed in executing the action with exactly the same behavioural instantiation every time for it to be a mimeomorphic action; it is what we intend that counts. But an action can fail. Mostly we will succeed so long as we do a reasonable job of behavioural mimicry; it does not have to be exact. The difficulty that humans find in exactly repeating behaviours is reflected in competitions like synchronized swimming or figure-skating – it is difficult for humans to execute complicated mimeomorphic actions.

The other kind of action is polimorphic (the word that current computers tend to try to 'correct' by changing it to 'polymorphic'), which means that the same action can be, and sometimes has to be, executed with different behaviours on different occasions, often in a context-sensitive way.[55] The prefix 'poli' is a combination of the Greek 'poly-', meaning many, and 'polis', meaning society, intended to reflect that to understand the appropriate behavioural instantiation of the same action requires one to understand the society. Compare the salute to the 'greeting'. Greeting someone in non-ritualized social groups can be done in many different ways and must be done in different ways on different occasions if the greeting is to retain its friendly meaning. If I use exactly the same form of greeting to my wife or partner every time we meet it will soon cease to work as a greeting and become a joke or an insult. I remember returning once from a long foreign trip and greeting my partner with 'you bastard'. We both immediately understood this to imply 'you have caused me a huge amount of pain because I have missed you so much'. Yet in any dictionary or any set of rules for speaking English, 'you bastard' would count as an insult. That is what's wrong with Watson and its use of the vernacular – Watson cannot (or has not) understood that the social situation determines the appropriate behavioural instantiation for a repetition of a polimorphic action. Watson cannot recognize the social situations in which it finds itself because it does not understand the society it is trying to analyse.

Yoshua Bengio, a pioneer of deep learning, writes in respect of this passage which I sent to him, in the context of an earlier conversation:

> I did not mean that this kind of phrase will be really understood any time soon, but that the scientific building blocks which we have developed around neural networks and deep learning make it plausible (for me at least) to believe that we will in a foreseeable future build machines which can

understand such a phrase in its context (which includes a lot of the history between you and your partner). [56]

I, in contrast, cannot envisage this coming about by incremental steps from where we are now because, among other things, where we are now does not seem to lead on to machines knowing the history of the relationship between me and my partner.

Go back to our spaceship observing the aliens: if, in their world, the aliens engage in both mimeomorphic and polimorphic actions, how are we going to make sense of their behaviours? (Remember we have no access to their intentions and therefore no access to their *actions*, only access to their *behaviours*.) Sometimes, in their world, a behaviour which is the equivalent of the utterance 'you bastard' will be used as an insult and sometimes as a deeply emotional greeting; we might be able to make sense of the equivalent of saluting in their world but we won't be able to make sense of the equivalent of greeting.[57] That is why both sociologists and intelligent machines have to be socialized if they are to make sense of, and mesh into, human societies; once more, what they have to learn is the language – learn it to the extent that they can make the same judgements as the native members as a matter of course, because language is how culture is transmitted.[58]

The body and artificial intelligence

This book concentrates on language and language-handling but it cannot entirely avoid the discussion of bodily practices. There is a purely scholarly reason for this: the pioneering critique of artificial intelligence, developed in the 1960s and 1970s by the late Hubert Dreyfus, turned on computers' lack of bodies. The next few pages will, therefore, refer to published works and the ideas of well-known scholars a little more heavily than the rest of the book. Dreyfus's critique is still important in philosophical circles. Furthermore, though language-handling is a fully adequate indicator of human intelligence, the development of human language as a whole depends on the human body. The development of language as a whole is different from the acquisition of language and fluency by an individual embedded in a society already equipped with a language. The form of language depends on the typical practices in that society but the individual's acquisition of language does not. The vital distinction between individuals' acquisition of languages and the collective development of languages is little understood. Let us begin with Dreyfus's influential and heroic critique of AI.

In 1967, Hubert Dreyfus, then a professor at Massachusetts Institute of Technology (MIT), published a notorious article called 'Why Computers Must Have Bodies in Order to be Intelligent'. He argued that our understanding of the world is tied to our bodily activities and therefore could not be reproduced by stationary boxes full of transistors. I have heard that as a result of this and his other criticisms of artificial intelligence Dreyfus had difficulty gaining tenure at MIT; the MIT Artificial Intelligence Laboratory was flying high at that time.[59] Consequently Dreyfus moved to Berkeley. In 1972 he published *What Computers Can't Do*, which developed his critique of AI at book length. Nowadays, this book and its 1991 second edition, *What Computers Still Can't Do*, still seem excellent, superbly written and worked-out critiques of AI, even if they are becoming dated in parts.

Many of Dreyfus's arguments still bear scrutiny, but we can now see that what was missing was the collective nature of human knowledge. The collective nature of human knowledge was not properly understood until the development of the sociology of scientific knowledge in the early 1970s – though the seeds of it were already to be found in the later philosophy of Ludwig Wittgenstein (for example, his 1953 book) and, if anyone relevant had known about it, the work of the medical research scientist Ludwik Fleck (first published in German in 1935 and not translated into English until 1979). None of these people were writing about artificial intelligence, of course; they were writing about the nature of knowledge. Wittgenstein was showing that the meaning of words is located not in the dictionary, but in the way we live. Thomas Kuhn, in 1962, put forward his famous theory of scientific 'paradigms', under which the way scientists act, think and interpret experimental findings is a matter of embedding in the social group of scientists in which they work, an idea to which, thirty years before, Fleck had given the name 'thought collective'. The important breakthrough was to show that this understanding of knowledge applied even to scientific knowledge which, up to that time, had been thought of as a domain entirely separate from ordinary thinking, acting and living. Only when it was realized that even scientific knowledge was embedded in society could this way of thinking about knowledge come to be seen as general. Up to then, any scientist whose ideas were affected by the surrounding society as opposed to being a pure reflection of Nature was thought of as suffering from a remediable flaw. We can now see that Kuhn's (and Fleck's) idea was Wittgenstein's 'forms of life' concept applied to the scientific community.

Anticipating Kuhn, the implication of the 'later Wittgenstein's' ideas was nicely explained in 1958, in terms of the germ theory of disease, by the Wittgensteinian philosopher Peter Winch. Winch explains that the discovery of a new germ is just an extension of existing scientific theory ('normal science' in Kuhn's terms), whereas the initial discovery of the germ theory of disease is integral with an entire change in the 'form of life' of surgeons (a 'revolutionary' change in Kuhn's terms). Nowadays, when we see surgeons ritualistically scrubbing we can think of their actions as confirming the existence of germs even as they think of themselves as removing germs from their hands. This thinking was much more fully worked out and applied to a number of concrete contemporaneous cases by the sociologists of scientific knowledge.[60]

Going back to what we can call the 'embodiment thesis' – the thesis about the importance of the body as championed by Dreyfus et al. – it arises out of a very different philosophical tradition. Dreyfus, in particular, was an expert on the philosopher Heidegger. Such 'phenomenological' philosophers take intense reflection on the way 'we' encounter the world as their starting point. Here, for example, is Wikipedia's account of Merleau-Ponty, a philosopher who is central to this tradition:

> Merleau-Ponty emphasized the body as the primary site of knowing the world, a corrective to the long philosophical tradition of placing consciousness as the source of knowledge, and maintained that the body and that which it perceived could not be disentangled from each other. (<https://en.wikipedia.org/wiki/Maurice_Merleau-Ponty>)[61]

Note that the phenomenologists differ from other traditions that deal with human thought in that they make the body central to cognitive life, and one can immediately understand the source of Dreyfus's critique: a stationary box filled with silicon-based components cannot be intelligent because it has no body through which to engage with the world in the way that humans engage. One can also see that phenomenological thinking has something in common with the Wittgensteinian approach because that approach too stresses the practical element of cognitive life – concepts and practice cannot be separated – as with the surgeon's scrubbing. The difference is that, for the phenomenologists, the practice is to be understood through the individual body's engagement with the world, whereas for the 'Wittgensteinians' the engagement is through 'forms-of-life', or 'thought collectives', or 'paradigms', a kind of engagement that emerges from the way of being in the world of social groups.

The individual versus group distinction is absent from the phenomenological analysis; for phenomenologists such as Dreyfus there are only individuals and their typical bodies.[62] The argument with artificial-intelligence believers concerning bodies was very nicely brought out by the intervention of Doug Lenat. Lenat pointed out that according to Oliver Sacks, one of his case studies, a woman called Madeleine who was profoundly disabled from birth, was nevertheless completely fluent in spoken language. Madeleine had acquired this fluency purely from conversational interchange with her carers, without any bodily engagement with the world.[63]

The critique of the need for bodily engagement, based on Madeleine, fits exactly with the idea of interactional expertise and with the individual side of the individual versus group problem. To acquire language from the group surrounding her, Madeleine needs only the mind in the machine. But *a group of Madeleines* (here we go beyond Lenat) could *not*, on their own, develop the human-like language that she acquires because developing that language in the first place does require bodily engagement with the world – it depends on the meat in the machine, not just the mind.[64]

Thus, although experiments we have conducted show that tennis can be understood by the blind even though they cannot play it, there would be no tennis in a *world* of blind people where nobody could play it.[65] If everyone was blind our conceptual world would be very different – something beautifully illustrated in H. G. Wells's short story, 'The Country of the Blind'.[66] In a world of blind people there would be no word for tennis; nor for a lot of other things familiar to us. So the language would be different, along with the concepts (this goes back to the idea of aliens on another planet, who we can observe but cannot understand). At the collective level, then, the conceptual structure of our language – the very substance of our language – depends on the things we do with our bodies, but this does not apply to individuals. To repeat, at the individual level what is crucial is 'the mind in the machine', but at the collective level where language is formed 'the meat in the machine' is equally important! Ironically, this is what makes a disembodied deep mind a candidate to be an intelligent machine – deep minds depend on the concept of interactional expertise to make them credible in the absence of interacting bodies!

Going back to the Introduction, we can again see why the Blofeld model of computer domination is puzzling: there's no meat in the machine. The Blofeld model, involving remodelled bathrooms, fast cars, swimming pools, cat-stroking and so forth, depends on the things humans do with their bodies. Why would a society of computers want nice bathrooms if they could not take a bath? Why would they crave power and wealth with no mates to attract or gods to satisfy? But individuals learn language and the physical understandings embedded in it, not from engagement with the world but from the groups in which they are brought up. So that resolves the Blofeld puzzle, as long as the computers are not autonomous but learn from (a certain type of) human language. Children are born and brought up in the midst of language-speaking groups and they learn the language of the group in which they are embedded. That, to repeat, is why individuals like Madeleine, who have bodies that are different from the mass of group members in various ways, can still learn the language of the group even though it refers to various kinds of activities in which they cannot engage. We now have enough concepts to develop a more nuanced understanding of what we really mean by artificial intelligence and its many variants. As we will see in the next chapter, it has six levels.

5

Six Levels of Artificial Intelligence

Now that we understand more about the interaction of intelligent machines, society and the body, we should be able to see more clearly what it means to say that a machine, or a group of machines, is intelligent – remember the collective problem is different from the individual problem. There are various meanings, set out in Table 5.1 as six ascending levels. They are divided into three groups of two because the biggest steps lie at the boundaries between the pairs, while subtle arguments characterize the differences within the groups of two (at least, in respect of the first two groups). The main focus of this book is on the transition from Level II to Level III, but the other levels are included to complete the picture and to reduce the possibility of confusion between Level III and the other levels. We'll see that the danger of the Singularity, if it is real, is most likely to apply at Level V and this has certain consequences in respect of what kinds of machines we should be trying to build. Our examination of the difference between Level III and Level IV will be especially useful for understanding what it means for computers to be world champions at chess and Go, and to explore the meaning of the computer 'Watson's' winning at *Jeopardy!* It will turn out that these successes fit better when located much earlier in the sequence of artificial intelligences. There is also an important argument about whether computer success like this actually *reveals* the way the human brain works, putting an interesting complexion on the relationships between some of these. We have little to say about Level VI but it is there for completeness and the contrast with Level V. I don't think it is possible to talk sensibly of the positive potential and the potential danger of artificial intelligence without understanding how these things differ from level to level. There is no single AI; there are a series of AIs.

Level I of artificial intelligence: Engineered intelligence

The first level is engineered intelligence which we already live with. Bear in mind that it can be argued that a simple thermostat is intelligent.[67] Engineered intelligences control your washing machine, your car, your power station and power grid and your missile launchers. Engineered devices are not generally exposed to Turing Tests, they make no demands in respect of bodily instantiation, there is usually no claim that they reason in human-like ways, their socio-economic impact is generally positive as they enhance human abilities, but they sometimes fail, occasionally disastrously, and sometimes they control us in undesired ways, though not intentionally. A characteristic over-interpretation of what has been achieved at this level is to take it that it is the first step to constructing artificial humans. They are no more the first step than a pocket calculator or a tractor is a first step.

I say that, usually, no claim is made that human-like

reasoning is being reproduced at this level, but this was probably not the case in the early days when intelligence was first being explored through machines; ambitious interpretations of their meaning are often associated with new technological developments.

Level II of artificial intelligence: Asymmetrical prostheses

There is a big overlap between Levels I and II because of the enormous force of humans' tendency to repair and the seductive power of anthropomorphism. If I can think of my car as having a personality, I can think of lots of things found at Level I as intelligent machines that fit into society where humans once fitted; one is tempted to say one has moved to Level II when it becomes easier to anthropomorphize, as it is with speaking computers such as Siri. Such things are 'asymmetric prostheses' and the distinction between Level I and II depends on little more than how the devices were intended to function and how they are used and thought about.

The term 'prosthesis' is used to indicate that a machine takes the place of human intelligence just as an artificial leg takes the place of a real leg or an artificial heart takes the place of a real heart. AIs are 'social prostheses' – they take the place of some human activity, not by replacing a bit of the body but by replacing a bit of society. As intimated, even tractors can be thought of as social prosthesis, a tractor taking the place of lots of humans with horse-drawn ploughs – but the term is more illustrative when applied to AIs such as your spellchecker, which tries to take the place of a human editor, or your airline's website which tries to take the place of a booking clerk.

The point about a prosthesis is that it does not have to

Table 5.1 Six levels of artificial intelligence

	Turing Test	Body	Human-like reasoning	Social/economic impact	Singularity
I Engineered Intelligence	N/A	N/A	No	Enhanced productivity	Accidental/malicious destruction (A/MD) [e.g. by hackers]
II Asymmetrical prostheses	May pass non-demanding	N/A	No	Enhanced productivity	(A/MD)
III Symmetrical culture-consumers	Pass demanding	N/A	Brute strength	Enhanced productivity and maybe social life	(A/MD)
IV Humanity-challenging culture-consumers	Pass demanding	N/A	Yes	Enhanced productivity and maybe social life	Individual geniuses
V Autonomous human-like society	Pass demanding	Human-like	Yes	Dangerous	Blofeld scenario
VI Autonomous alien society	N/A	Alien	No	Don't know	Don't know

work in exactly the same way or even do exactly the same job as the thing it replaces in order to be satisfactory. Artificial legs don't do exactly the same job as real legs, but we adjust the things we do with the rest of our bodies to make up for most of the differences. Artificial hearts have many different characteristics from real hearts but, again, our bodies adjust to make do. And you usually fix your spellchecker's mistakes without noticing that they have been fixed by you – probably thinking the spellchecker is doing the same job as a human editor. Likewise, you put up with all the awkwardness and dangers of a non-human interface when you make an online booking for an airline ticket as though it was something ordinary.

What we do with all these things – with prostheses – is, of course, repair! Indeed, our capacities for repair are so deep and profound, since we have to use them all the time just to understand the speech of other human beings, that we 'over-repair'. This happens when we anthropomorphize our cars and animals and make good the mistakes of con-artists and bogus doctors. The danger is that we are so good at repair that we don't even notice ourselves doing it and so mistake machines' abilities for human-like abilities. This also causes some believers in artificial intelligence to make over-optimistic claims for the technology – as when they say the Turing Test is too easy and has already been passed – and it leads users of the technology to put too much faith in what it can do, the problems becoming clear only when circumstances require rule-breaking behaviour or the understanding of rule-breaking behaviour. We call Level II *asymmetric* prosthesis because we can, and continually do, repair the machines' faults, but they cannot repair ours.

The characteristic over-interpretation at this level, and it is an important and widespread malaise, is to think that an asymmetric prosthesis is a *symmetric* prosthesis. This happens because the amount of repair and anthropomorphism that has gone on when an asymmetric prosthesis does good work is normally invisible. A determined effort is always needed to convert an automatic aid that has become increasingly familiar into something strange once more; only then do the normally ignored human inputs become visible. The problem is confounded because today's asymmetric prostheses can do quite a bit of repair on their own using precedent-based, statistical methods. Siri, Cortana and the rest make a pretty good job of appearing to make sense of what you say, even when you don't say it with utter clarity. But the distinction between this and full symmetry is vital if we are to understand artificial intelligence.

Level III of artificial intelligence: Symmetrical culture-consumers

From Level II to Level III is a huge step. Level III, if it ever happens, will be the era of fully symmetrical prostheses – social prostheses that are so good at repairing our broken speech and other rule-breaking activities, and so good at recognizing and absorbing our precedent-setting activities, that they can respond appropriately to even the most novel interactions and recognize when they are legitimate. Only at Level III will machines pass the best-designed, extended and demanding Turing Tests. At Level III, computers will be as fluent as we are and this means they will be able to engage in polimorphic actions. They will have absorbed our cultures by one method or another and will understand the shifting nuances of the social contexts in which they are embedded. Even if the method they use differs from the way humans execute their intelligence, from the outside it will be impossible to see the difference. From the outside they will appear fully to share our human culture (or some local variant of it), so somehow that culture must have been absorbed into them in the same way that an anthropologist or an interactional expert absorbs the culture of those he or she is studying. Ava, Samantha and HAL, as they are portrayed, are (at least) Level III artificial intelligences; they are thoroughly accomplished culture-consumers and have drunk in the human culture around them well enough to reproduce it in conversation which is indistinguishable from that of their human companions.[68] Ava, Samantha and HAL may be psychopaths, but so are some humans and, as with humans, you don't find such things out until some way down the line; in their day-to-day interactions through the medium of speech they are indistinguishable from Westernized humans.

The step from Level II to Level III is enormous, but the size of the step can be illustrated by very simple examples, such as the repair, or non-repair, of rule-breaking and precedent-setting creativity (for example, the non-repair of 'wierd' as used in this book). Passing a truly demanding Turing Test will indicate transition from Level II to Level III.

Level IV of artificial intelligence: Humanity-challenging culture-consumers

The difference between Level III and Level IV is a subtle one. Let us suppose we have learned how to build Level III devices so that in everyday operation they are indistinguishable from humans and just as fluent – they pass every Turing Test, even the most demanding we can dream up. The question that still remains is whether those devices' internal workings are the same as those of humans. This question is not of huge interest to me as a sociologist; those who endorse the way of thinking that drives this book will have to admit that the 'socialization critique' of AI has been overcome if the difference in the everyday functioning and fluency of humans and machines becomes impossible to uncover in the face of maximally demanding Turing Tests. But those of a more metaphysical inclination will still be interested in whether human abilities are being reproduced and we will all be interested in whether we have now proved that we are just meat machines and humanity doesn't have a soul, nor anything unique that could stand in for it; it is hard for even the most sociologically inclined not to be at least a little interested in this question. Here, Searle's intended critique of AI via the Chinese Room would displace the way the thought experiment has been used in this book, and the claim that the internal states of humans and computers was of no interest would no longer be true; the question now would be to understand not just the performance of computers, but what underlies computers' performance.

One must not make the answer a truism by insisting that doing things like humans means using the same *biological* mechanisms, otherwise the problem would become not one of reproducing human abilities but reproducing humans. So we have to accept that thinking 'like' a human, while using silicon chips or some such, potentially meets the criterion of Level IV – reproducing human internal states (though there are nice philosophical questions here).

Post-anthropomorphism

A new term is needed to describe the possible potential of machines if we are to avoid self-serving rhetoric. Too often in the field of AI, words that describe human abilities are used to describe programming methods; the terminology confounds the analysis of the difference between what computers do and what humans do. The computers seem to be doing what humans do by definition: they 'learn', they 'think', they 'decide' and so forth, prior to any analysis of what they actually do. In this way, the term 'deep learning' could appear to have solved the problem of human learning through anthropomorphic definition even before the process has been investigated.

The equal and opposite side of persuasively defining the abilities of computers through anthropomorphism is that it is all too easy for critics to argue that the enthusiasts are simply being anthropomorphic every time it is said that a computer has accomplished some human-like feat. That way, also by definition, no computer will ever reproduce human abilities until it is an exact reproduction of a human in physical constitution as well as everything else.

What we need to ask is whether some computer or program has become so human-like that it no longer makes sense to talk about anthropomorphism; it makes no more sense than to use the charge of anthropomorphism when talking of the abilities of another human being. It would be as though some early explorers encountered a tribe in a distant land and some of their party talked of them as learning and thinking while another faction complained that this was anthropomorphism since they weren't humans like us. There comes a point where the charge of anthropomorphism becomes unacceptable prejudice. If the ambitions of the computer enthusiasts come to fruition, then, in respect of computers, we will have entered the 'post-anthropomorphic' age. We'll use the history of computerized chess and Go to explore what post-anthropomorphic might mean. The discussion serves a second purpose – to show that the success of computers in these board games does not tell us as much about human intelligence as has sometimes been said.

Chess, Go and anthropomorphism

Hubert Dreyfus famously claimed that no computer would ever beat a chess grandmaster because it was impossible to set out and program the rules used by grandmasters. In his 1972 book *What Computers Can't Do*, repeating it in the follow-up book twenty years later, he wrote:

> In chess programs . . . it is beginning to be clear that adding more and more specific bits of chess knowledge to plausible move generators, finally bogs down in too many ad hoc subroutines. . . . What is needed is something which corresponds to the master's way of seeing the board as having promising and threatening areas. (1972: 296)

But Dreyfus was wrong about what was necessary for chess programs to be successful; it turned out computers did not need 'the master's way of seeing the board'. It turned out, I think to almost everyone's surprise, that more power – more brute strength – could do the job of winning at chess, and Deep Blue, a program developed by IBM, did beat grandmasters. Such programs depend on estimating the value of a move by reference to a set of 'heuristics' that can provide a measure of the power of a position on the board, but, more importantly, they have the ability to calculate ahead further than humans can calculate. When humans play chess they calculate: 'if I do this my opponent could do this or this, then I could do that or that, and I will be in a better/worse position than before whereas if I do that my opponent . . . etc.'. This move-tree expands exponentially – it explodes – and we know that to calculate this way from the beginning through all possible positions to the end of the game is completely impossible: it would need a computer many times bigger than the universe. That is why it was thought for a long time that computers would not be able to beat grandmasters, since to play chess well required some intuitive sense of the state of the board. But powerful computers can calculate a few extra moves ahead compared to humans and this, it turned out, was enough to enable them to win, and we now know that powerful-enough computers will always beat humans at chess. Thus, in 1997, Dreyfus was to say:

> I said that a chess master looks at only a few hundred plausible moves at most, and that the AI people would not be able to make a program, as they were trying to do in the 60s and 70s, that played chess by simulating this ability. I still hold that nothing I wrote or said on the subject of chess was wrong. The question of massive brute-force calculation as a way of making game playing programs was not part of the discussion, and heuristic programs without brute force did seem to need, and still seem to need, more than explicit facts and rules to make them play better than amateur chess. But I grant you that, given my views, I had no right to talk of necessity. (From a debate between Daniel Dennet and Hubert Dreyfus, *Slate* May 1997[69])

This is the kind of thing that keeps taking the critics by surprise; we, on the right-hand side of Table 2.2 (p. 30), are just not very good at foreseeing what the siliconeers and their colleagues on the left-hand side will do next, and what sheer power and ingenious work-arounds will produce. Both sides are discovering what brute force and, nowadays, precedent-based methods can accomplish.

But the chess example showed that the victorious chess-computers were not playing in the way that humans play, so success through brute force is not the same as creating a machine that plays like a human. There are subtle distinctions at work here: a chess program like Deep Blue can beat a human at chess but this, it turns out, is to say nothing more portentous than that a tractor can beat a human at ploughing. The current state of the argument is that computers play a recognizably different game from humans – at least a different game from that played by grandmasters in the past. Consider this description of the current Norwegian world chess champion:

> On the other hand his middle game and endgame playing resemble how engines play chess. Humans struggle to, for example, return a piece to the position it was located one or two moves earlier, even though it would be the objectively best move. Computers don't care about the past and play the move that their calculations determine is the best. Carlsen seems to be able to avoid this human bias, and play more like a computer. Magnus Carlsen has not only become the World Chess Champion, he has created a different style of playing. (<http://theconversation.com/how-computers-changed-chess-20772>)

If it is the case that Carlsen, the human, now plays like a computer, then the difference between the machine and the human once more disappears but for uninteresting, even perilous, reasons – perilous because it is a step on the way to the Surrender: we can always choose to act like machines and engage in a growing preponderance of mimeomorphic actions in our lives. We have to understand, but try to resist, humans choosing more and more to act like computers, thus erasing such distinctions in style. Nevertheless, the fact that, historically, they acted differently is enough to show us that, insofar as chess is concerned, if the equivalent of the Turing Test is being passed, it is only because we choose not to ask the right kind of difficult questions of chess. Having humans become like machines cannot and must not be the solution to the problems of artificial intelligence.

Having brought the Turing Test into the argument again, let us note one further thing about chess, and this also applies to Go, which we will go on to discuss in the next paragraphs. I say 'insofar as chess is concerned, if the equivalent of the Turing Test is being passed, it is only because we choose not to ask the right kind of difficult questions of chess', but what would the right kind of difficult questions be? Following the Carlsen example, they would be questions about style – one could imagine a grandmaster playing against a computer and a non-Carlsen-like human, and recognizing the human style of play. But we could not use this test to answer the question of whether the chess computer had made the transition from Level II to Level III because there would be no way of testing for the computer's ability to repair faulty language in a context-sensitive way. One cannot play chess or Go in a mumbling, repair-needing way because the inputs are digitized by the way the board interacts with the pieces. So we should not really be dealing with chess computers and Go computers in this section; we should have dealt with them under Level I or, at best (given our propensity to anthropomorphize), Level II. Chess and Go don't really enter the race when it comes to mimicking the range of human abilities that are *really* hard to reproduce mechanically.

The chess example illustrates one more very important thing to be said about humans' relationship to computers and the Surrender. Sometimes it is good to surrender our human creativity and inventiveness to machines that are dumber than we are because there are times when dumb machines do a better job than clever humans. Chess machines play better than humans in terms of winning, but that does not tempt us to make the switch. But anywhere that there is no value in the creativity associated with human activity is a potential location for replacement of humans by machines: the Industrial Revolution consists of such examples. A more contentious location that depends on 'intelligence', rather than energy, strength, endurance and unwavering vigilance, is car-driving. About 3,250 people die every day on the roads worldwide. That's equivalent to about ten jumbo jets crashing. And about twenty-five times as many people as die, are injured, often disabled. It is remarkable how few people know these statistics, though everyone knows about the one death caused by a self-driving Tesla car. Car-driving is a lot of fun and provides many opportunities for young people to show off even as they kill and maim; furthermore, it is possible to imagine circumstances where a machine could never drive as effectively as a human: say someone is bleeding to death and I am driving them to hospital: encountering a traffic jam, I decide to drive across the front gardens of a series of houses, knocking down their fences, so as to skirt the stationary cars; I can't imagine the program that could choose this potentially life-saving action. So we will lose a lot when all cars become self-driving. But what we will lose is far outweighed by what we will gain, even though driving will become dull and unimaginative: boring automatic driving in convoys is going to be much safer on average than what we currently have. Where such situations occur, the case has to be argued and we should accept the Surrender of our rights and flair. Whether we will actually accept this or not is another matter, as the argument about guns demonstrates: the number of people dying from gunshots in the USA is not dissimilar to the number dying from road accidents! But notice, that this argument is not about the ability of humans to mimic machines; it is about the uses of Levels I and II AI engineering.

Now let us get back to the main theme and look at the program AlphaGo and its success at becoming world champion at the board game Go. This was not a matter of its calculating a few more steps ahead than humans but, as I understand, of its recognizing patterns on the board in the way a Go master sees them, in the same way that Dreyfus said a chess grandmaster sees chess patterns. I would imagine that a chess program based on the same principles as AlphaGo would satisfy all Dreyfus's strictures on what it is to play chess like a human.[70] Does this mean that AlphaGo has reached Level IV of artificial intelligence?

To see more clearly what is at stake, consider Dennis Hassabis's description of AlphaGo – the deep-learning program his team developed that 'taught itself' to become the world Go champion. AlphaGo developed ways of playing that were previously unknown to humans and which cannot be explicated by AlphaGo or its creators. Hassabis, speaking of AlphaGo, remarked:

> Now I just want to take a moment to discuss intuition and creativity which I think AlphaGo has demonstrated here. I, sort of, use the word intuition quite a lot but what do I mean by it? I think one useful way to think of intuition is that it is implicit knowledge that has been acquired through experience but it can't be consciously expressed or communicated.[71]

The question is whether Hassabis is correct to say that AlphaGo has 'intuition', or implicit knowledge (or 'tacit knowledge' if he had used that term), or whether he was using the terms in an anthropomorphic way. Can we, then, use this discussion to refine the meaning of 'post-anthropomorphism'?[72] The definition developed here is based on the central thesis of the book: human-like computers will have to be as context-sensitive to the society around them as humans. Since, I am going to argue, the right kind of Turing Test can distinguish machines with that kind of context-sensitivity from those without, the right kind of Turing Test will tell us when the age of post-anthropomorphism has been reached.

So how about AlphaGo? It seems to play in a way that is indistinguishable from a human, except better, and, as Hassabis says, it seems to use intuition and creativity or its machine equivalent. Falling back on philosophical argument from an earlier book, it could be said that AlphaGo could, at best, be *mimicking* human actions since it has no intentions and intentions are integral to actions.[73] But maybe AlphGo has intentions! As explained earlier, I am suspicious of strong claims about internal states. So let it be the case for argument's sake that to say that AlphaGo uses intuition when it plays Go is *not* anthropomorphic. That sounds like quite a big admission and it would be if it meant that AlphaGo had made a significant step to using intuition or the like in a range of activities. But our deeper argument remains that not a lot turns on the admission because of what AlphaGo has *not* achieved. Its accomplishments, intuition-based or not, lie in the realm of what we will call 'technically defined goals' rather than 'culturally defined goals'. We have already hinted at this when we discussed what a Turing Test for a machine like AlphaGo would imply: it would not involve testing the machine's abilities to repair human mistakes and would lead us to locate such a machine at Level I.

The board game Go (like chess) is a constrained formal environment with a well-defined end-point. To learn to win at Go does not require any embedding in, and learning from, human culture once the rules and aims have been programmed in. There is no social context that people worry about when AlphaGo plays except that one imagines that if there was nuclear war, or a worldwide epidemic or some such, human players would stop playing while the computer would carry on regardless. But if we do not consider these outside environmental factors part of the game, then we can understand how it is that AlphaGo could learn to win at Go by playing millions of games *against itself*, whereas neither such a program (nor any human) could become, say, fluent in English by holding any number of conversations *with itself*. To become fluent in English does require contextual understanding and therefore does require embedding in the culture – one must talk not to oneself but to all the other language speakers if one is to come to speak like them.

Technically defined goals are pretty well fixed once they have been defined – the meaning of winning at chess or winning at Go or winning at snooker are all pretty well fixed, whereas the meaning of fluency in a language is not fixed. I say these things are 'pretty well fixed' because there is always a chance of them changing, as illustrated by a very old newspaper cartoon by Bill Tidy. The cartoon shows a snooker game in which one player has reached up and grabbed the edge of the overhead light and swung from it to hit his opponent squarely on the chin with both feet, an onlooker remarking, 'when he was younger he would have seen that coming'. Of course, the rules have been broken but we can imagine them changing to allow such a move if our society was to change radically.[74] But, setting this kind of possibility aside, technically defined goals are fixed. Culturally defined goals, like fluency in the language, are always changing as society frequently redefines the nature of the language – all kinds of new usages are continually entering natural languages; this is the reason one has to engage with the culture to remain fluent.[75]

Going back to intentions and actions, the crucial kind of action that signifies human abilities is polimorphic action, while AlphaGo, even if we speak of it as having intuition, is executing mimeomorphic actions, albeit a complicated kind of disjunctive mimeomorphic actions.[76] Technically defined goals are a domain of mimeomorphic actions and do not turn on sensitivity to social context, whereas culturally defined goals are a domain of polimorphic actions, which do depend on context sensitivity. Since we are tying post-

anthropomorphism to context sensitivity – and the Turing
Test – AlphaGo's astonishing accomplishments are not yet
there, even if one does allow it to be said that intuition is
being displayed.

The body once more

Notice that neither Level III nor Level IV require that the
computers that pass the Turing Test have human-like bodies.
This is because we are talking of individual machines and we
know from the interactional expertise argument that individuals can absorb culture from human societies without having
human-like bodies – which is, effectively, what sociologists
do when they explore alien societies. We can think of those
acquiring interactional expertise as parasites; let us think of
them as a kind of beneficial parasite. I say the sociologists do
it without having human-like bodies because they (usually)
do it without using their bodies in the same way as they are
used in alien societies – the sociologists are the equivalent
of a person who has always been in a wheelchair learning
about tennis; one should say the sociologists are *effectively*
without human bodies. To repeat, the crucial point about
those without bodies, or effective bodies, whether machine or
human, is that neither could autonomously *create* human-like
culture unless they had human-like bodies.

   This point is worth reiterating. When I, the sociologist,
am learning a new culture, such as that of gravitational-wave
physics, I am in a similar position to that of a Level III or
IV AI which is trying to absorb the culture of a sub-group
of human society. I claim that I can succeed to the extent
that I tend to make similar practical judgements to those
who actually practise the activities pertaining to the culture,
even though I am effectively disembodied in respect of those
activities. But even after I have acquired that culture I am

still in the parasitical state of the individual as opposed to the autonomous state of the embodied collectivity. The interactional expert is a parasite drawing the cultural blood out of societies that have already been established by contributory (i.e. embodied) collectivities of experts. I can make changes to those cultures by contributing good judgements, but I could not invent a culture/practice *de novo* because to do that you need the ability to practise and to have the goodness of those practices affirmed by other practitioners. I also could not go off with a group of my 'disembodied' buddies (a group of other sociologists of science with interactional expertise in gravitational-wave physics) and push the physics forward because our understanding would drift away from the understanding of those who are practising – our interactional expertise would soon transform itself into something like the language spoken by the members of a cargo-cult – for example, the Pacific islanders who hosted the American forces during the Second World War. The cargo-cultists, or so the story goes, learned to associate aeroplanes with beneficial cargo, and when the Americans left continued to worship aeroplane-like icons and so forth in the hope of acquiring more goods. The cargo-cultists also continued to use terms common to the donor society, but with changed meanings. That is why interactional experts remain parasites, even when they can contribute to practical judgements while they are immersed in the culture – take them away from the culture and, after a while, the meanings are likely to change or degrade.[77]

Does Jeopardy! *show that Level IV is being approached?*

To turn again to our iconic AI figure, Ray Kurzweil thinks that the current success of precedent-based AI is pointing to success not only at Level III, but also at Level IV, because the human brain simply *is* a precedent-based, pattern-recognizing device, like deep-learning programs. This argument is worth exploring further. Ever since it beat the champions at a US TV game called *Jeopardy!*, a lot of attention has been given to a pattern-recognition-based program called 'Watson', developed by IBM. In *Jeopardy!* contestants are given answers and have to guess the corresponding questions. To do this, they have to navigate their way through various puns and allusions. For example:

> A long tiresome speech delivered by a frothy pie filling.
> A garment worn by a child, perhaps aboard an operatic ship.
> It can mean to develop gradually in the mind or to carry during pregnancy.

The answers are: 'meringue/harangue', 'pinafore' and 'gestate'. Watson found the solutions and performed better than the best humans. This is another one of those surprises that the new artificial intelligence has delivered and that could not have been foreseen a decade or so back, even though it uses an assembly of 'tricks'.

Kuzweil (2012) explains how Watson works: '[The] machine zeros in on key words in a clue, then combs its memory . . . for clusters of associations with these words' (p. 157). Kurzweil tells us that its memory bank is built up from Watson reading:

> hundreds of millions of pages on the Web . . . Ultimately machines will be able to master all the knowledge on the Web – which is essentially all of the knowledge of our human-machine civilization. (p. 159)

> [Watson] obtained that knowledge by actually reading 200 million pages of natural language documents, including all of Wikipedia and other encyclopedias. (p. 166)

He tells us:

> It rigorously checks the top hits against all the contextual information it can muster: the category name; the kind of answer being sought; the time, the place, and gender hinted at in the clue; and so on. (p. 158)

> It contains hundreds of interacting subsystems, and each of these is considering millions of competing hypotheses at the same time . . . Doing a thorough analysis – after the fact – of Watson's deliberations for a single three-second query would take a human centuries. (p. 160)

> Watson can understand and respond to questions based on 200 million pages – in three seconds! (p. 166)

Kurzweil anticipates systems that will scan everything on the web and there is no reason to doubt that this will come about and, as intimated, for the sake of argument – trying here to get our imaginations around exponential growth – we can anticipate still bigger systems that will also scan and transcribe every TV and radio transmission and every conversation that takes place within range of a microphone in a device that transmits via electromagnetic waves.[78] Would this make Watson human-like?

At a trivial level the answer is 'no'. Humans do not have this kind of access or programming speed – 'three seconds versus hundreds of years'.[79] But for the sake of argument, let us not worry about it and give Watson the benefit of the doubt since it plays better than humans and that could be where all the processing power is going. Instead, let us go back to another intriguing claim of Kurzweil's:

> Some observers have complained that Watson does not really 'understand' the *Jeopardy!* queries or the encyclopedias it has

read because it is just engaging in 'statistical analysis'. [But] the mathematical techniques that have evolved in the field of artificial intelligence (such as those used in Watson and Siri, the iPhone assistant) are mathematically very similar to the methods that biology evolved in the form of the neocortex. If understanding language and other phenomena through statistical analysis does not count as true understanding, then humans have no understanding either. (Kurzweil 2012: 7; he expands on this on pps 157ff.)

Here, then, Kurzweil is disputing the distinction between 'brute-force' methods, or precedent-based statistical methods used by the new artificial intelligence, and human intelligence. His claim is backed up by his model of the human brain; we will discuss it at length in the next chapters. This claim typifies the kinds of argument that are likely to take place as computers become increasingly successful at Levels II and, maybe, III of artificial intelligence; there are going to be arguments about whether Level II is really Level III and whether Level III is really Level IV – namely, that anything that performs as well as a human is, by that fact, demonstrating how the human brain works.

The little experiment on garbled and jumbled passages of prose can also be brought to bear on the tension between precedent-based statistical methods and human sense-making. Remember the two passages, the top one being very hard to decipher for humans, while the lower one is quite readable.

at aoccdrnig be olny rsceareh a it mtaetr in oerdr the waht dseno't to ltteres a are, the iproamtnt pclae Uinervtisy, taht tihng lsat is the and wrod Cmabrigde ltteer in the in rghit frsit.

aoccdrnig to a rsceareh at Cmabrigde Uinervtisy, it dseno't mtaetr in waht oerdr the ltteres in a wrod are, the olny

iproamtnt tihng is taht the frsit and lsat ltteer be in the rghit
pclae.

It seems to the human eye that the lower one is readable, or
repairable, because we can make sense of it whereas the top
one does not make sense. We are saying, in other words,
that we can read the lower passage because we have a pecu-
liarly human way of dealing with such things; our method of
repair depends on imputing sense to passages of speech or
writing – on *understanding* – not on doing statistical analysis
of precedents. We know we are not solving anagrams in the
way we showed a computer could solve them, or we would
be just as good at reading the top passage as the bottom one,
so it certainly is not *that* brute-force method that is in play
when we read the passage. But the possibility is still open
that what feels like-sense-making to us is really Kurzweil-
type pattern matching. Consider Table 5.2, which shows all
the phrases that make up the garbled passage, haphazardly
arranged.

Those phrases are pretty easy to make sense of and read,
so what we thought was making sense of the whole passage
could actually have been making sense phrase by phrase.
But such sense-making *could* be accomplished merely by our
recognizing these phrases – that is discovering close precedents and matching them up via statistical analysis just as the
computers do – because, in our lives, we have encountered
them many times before. This is not something that is plausible for the whole passage but is plausible for the individual
phrases. Thinking of it his way makes it at least possible that

Table 5.2 Phrases from the 'Cmabrigde Uinervtisy' passage

in the rghit pclae	in waht oerdr	the olny iproamtnt tihng
it dseno't mtaetr	Cmabrigde Uinervtisy	the ltteres in a word
the frsit and lsat ltteer	rsceareh at	aoccdrnig to

Kurzweil is right and what *feels like* sense-making is merely pattern-matching.[80]

To indicate some of the problems ahead for the kind of claims that increased success in artificial intelligence is likely to encourage, we need to note that there is also a level of meta-understanding involved in the 'Cmabrigde Uinervtisy' example: we humans know when to give up on the first completely garbled passage, and that requires not pattern-matching but higher-level understanding of the task in which we are involved. Even that first passage could probably be deciphered by a bright anagram-minded human given less than an hour's work, but somehow we understand enough about what is going on to know that this would miss the point. The rules of the game (and that includes the meta-rules) are implicit in the game and learned through social immersion; without even thinking about them we unconsciously know these rules include that, in this case, you shouldn't spend more than a couple of minutes on finding the solution, other-wise, for the appropriately context-sensitive human, it counts as 'insoluble'.

Level V of artificial intelligence:
Autonomous human-like societies

Level V is like Level III, or Level IV, except that the computers have human-like bodies and so can expect to have human-like intelligence, but they will go beyond the level of being individuals parasitical upon existing human societies. They will be capable of more than those whose expertise is interactional; interactional expertise is parasitical on human societies engaged in the practices that give rise to language, but theirs will be independent. We can imagine groups of these computers forming their own autonomous societies with similar motivations to humans. We can imagine their wanting remodelled bathrooms and ocean-going yachts. It is such computers that might give rise to the Ian Fleming inspired 'Silicon Spectre'. They would need to eat and reproduce sexually in a competitive way. But why would we make such things except as individual laboratory experiments – to show we could! There is, I suppose, a dystopian scenario where they are developed as colonies of slaves who then take their freedom by force. But if slaves are wanted, one would think Levels III or IV with purpose-built bodies would be better – artificial slaves with human-like-bodies and human-like desires are likely to be a real nuisance.

Level VI of artificial intelligence:
Autonomous alien societies

This level comprises intelligent machines with *non-human-like* bodies but capable of building more of themselves and improving on the design as successive generations unfold. Their desires and intentions will be opaque to us – as opaque as the intelligence of extraterrestrial aliens. But, precisely because of this, there is nothing to cause us to suppose that they will be power-mad in the way that humans are, and no reason to think they will enslave or destroy us even though they would have the capability – they may simply not want to do any such thing; maybe they'll just want to lie in the sun. We really have no idea what such a level would look like, or act like, but before we panic about the Singularity we need to think hard about Levels V and VI and what follows from the difference in embodiment.

Concluding remarks

Humans as a group get their desires, intentions and ways of living from, among other things, the way their bodies work. Individual super-intelligent computers that do not have human-like bodies will probably not have Westernized human-like desires unless we feed them in – unless we encourage these isolated computers to be parasites on those bits of our society who already believe that unlimited accumulation of consumer goods is the natural way of being. There is, therefore, no necessity for them to be power-mad; if they are, it is likely to be a result of *our* carelessness or maliciousness. Groups of super-intelligent computers with human-like bodies might well be power-mad, but why should we construct such things? From considering the levels of artificial intelligence in the context of the old AI arguments, we reach the same conclusions as were found in the Introduction – the chances of a Silicon Reich coming to pass seem slight, but it still seems unwise to build self-replicating intelligent machines except under very carefully quarantined circumstances.

Outside of science fiction it would seem odd to build sexually reproducing machines with human-like bodies that would likely share our desire for power. Best not do it beyond laboratory experiments with neutered robots, and maybe not at all. Nearly all the current interest, and certainly the main interest in this book, is in the transition between Level II and Level III, and transition to Level III is going to be a prerequisite to anything more ambitious. Given what my current spellchecker can't do, we don't seem to have got very far, even though lots of artificial-intelligence believers believe we are nearly there. We are still a long way from building machines that will pass a really demanding Turing Test.

Deep Learning

Precedent-Based, Pattern-Recognizing Computers

The developments that have taken place in artificial intelligence over the last decade or so have been remarkable and their consequences, as already intimated, have been mostly unforeseen. The next two chapters cover these developments and explain them in as simple a way as possible. There are some philosophical issues that have to be covered for completeness – notably the difference between bottom-up and top-down pattern recognition.

Extended Moore's Law

Once more, let us turn to our iconic AI enthusiast, Ray Kurzweil. Central to Kurzweil's 2005 book, *The Singularity is Near*, is what he calls 'The Law of Accelerating Returns', which is a more general form of what we will call 'Extended Moore's Law'. Moore's Law, which is based on an analysis of the history of the development of integrated chips, states that the number of transistors in an integrated circuit doubles approximately every two years. This explains the extraordinary increase in the power and capacity of computers and the extraordinary *acceleration in the rate of increase* of power. The doubling and redoubling in the same time period is known as 'exponential' growth. With exponential growth, what initially looks like steady progress soon changes into explosive acceleration. Kurzweil points out, correctly, it seems to me, that human minds find it hard to foresee the outcome of exponential processes; when asked to predict the future we tend to look backward and project the past steady-looking trend forward into the future, so we don't see the coming explosive period of growth. This is one reason why critics of AI fail to foresee the way things are going: we don't get our heads around exponential growth. That is why it was hard to see that what we thought of as certain intractable problems of AI, such as beating grandmasters at chess, could be solved by brute-force methods. Because the potential of brute-force methods grows along with the exponential growth in the power and capacity of computers it meant that grandmaster-beating chess programs, once the very frontier of computing, were soon to be found on laptop computers and even smartphones. At the time of Kurzweil's book, this was close to unimaginable.

Kurzweil argues that something likes Moore's Law applies to all technological processes. The length of time over which the rate of progress doubles is getting shorter at an exponential rate and that is his Law of Accelerating Returns. He deals with the obvious resource and environmental problems regarding 'the law' by invoking nano-technologies.

Here we look at computers alone. We'll take Kurzweil's word for it when it comes to increase in the rate of acceleration and call it Extended Moore's Law. Once more, we are going to assume every technical advance that the computer enthusiasts say will happen, *will* happen: we will ignore the logistic problems and act as naively as we can. Our concern is not to do with current technical limitations; what we are interested in is what these extraordinary advances still *won't* be able to bring about just by growing in power, however fast they do it.

Neural nets and their successors

The pattern- or precedent-recognizers that underlie the new AI are based on 'neural nets'. Neural nets have been around for a long time. They were invented in the late 1950s and there was a flourish of enthusiasm for them in the 1960s under the name of 'connectionism'. According to Kurzweil, in his 2012 book, this early initiative was beaten down by those who believed the way forward was step-by-step, 'symbolic' programming, with the result that connectionism died a premature death; he, encouraged by his mentor, Marvin Minsky, was among those persuaded that the connectionist approach had no future. But now things have turned around and huge advances are being made by neural nets and their successors. The reason is that, in a sense, neural nets create their own solutions to problems, rather than being guided by human-designed computer programs.

Neural nets run on digital computers, which means that the code to run the neural net is initially written by a human, but once written the program lives its own life, as it were; without a huge effort, exactly how the program is working can remain mysterious.[81] What all this means is that neural nets can learn to process better and better at a rate that is enormously faster than if they depended on humans to work out each new step. With this approach there no longer seems any obstacle to computers approaching the complexity of the brain – not in terms of the software, anyway, while Moore's Law takes care of the hardware.

Because the details of how neural nets work are not too important, just a brief outline with some imaginative pointers should be sufficient for the purpose of understanding the principle of how a machine can 'learn for itself'.

Neural nets run on ordinary digital computers so the following mentions of nodes refer to virtual nodes created by software. Roughly, digital code is designed to assign weights to the virtual links between virtual nodes in a virtual network. These weights affect what happens at either end of the link. Somehow these weights are caused to change by some kind of feedback loop, depending on whether the neural net produces the desired outcome or not, so the way signals travel through the network is continually changing; the machine is continually transforming itself into something different, like a child as it grows up. This is how the links between neurons in a human brain work – signals frequently travelling along certain paths make those paths better transmitters and vice versa, so the network connectivity is continually changing according to what is being experienced, with some experiences being reinforced and some weakening.[82]

Because both the strength of the weights and the arrangement of the network are continually changing, unpicking what has gone on in such a device would mean retracing every modification in the whole network since it was switched on. The logistical impossibility of this lends such devices a sense of mystery (though the setting at any one time can be recovered). This is one of the attractions of these devices – they seem to reproduce the mysteries of the developing brain, and to reflect the way neuronal connections form and reform in the brain, hence their name – 'neural nets'. In the next paragraph we present a 'just so' story which is, I hope, just good enough to give some sense of what happens without any attempt to describe the intricate technicalities.[83]

Suppose there is some input at one end of the neural net – say a TV camera that turns an image into a matrix of pixels represented by a series of numbers corresponding to the position and brightness of each pixel – and, say, each of these numbers is fed into a node. Suppose what you present to the TV camera are images of capitalized letters of the alphabet and that the output at the far end of the neural net is constrained to print letters of the alphabet only: the system, note, has already been given some implicit training in restricting its outputs. Now suppose you present an 'A' at the input and what gets printed out is something like a 'P'. You have some way of giving the whole system a 'negative nudge' that tells it to change its weights a bit in a random way and you try again. This time, shown 'A', it produces something like 'W'. So you give it another negative nudge and try again. You keep on trying until it produces something like 'A' when shown 'A', and you apply some means of saying, 'make those weights a bit more resistant to change when you do your next set of adjustments' – in other words, you give it a 'positive nudge'. You continue until an 'A' emerges in response to an image of an 'A' at the input end. Effectively, you are 'training' the system with the electrical equivalent of punishments for failure and rewards for success. Having finished with 'A' you move on to 'B' and repeat the process, except that now the system is biased towards preserving those weights that were reinforced when it was learning 'A'; it is, to be anthropomorphic about it, 'remembering' how to recognize an 'A'. All this nudging is exhausting, so you write another program to automate the feedback. This program keeps trying different letters at the input side which it recognizes by their digital codes, and it knows the digital code of the outputs and automatically gives negative nudges when the codes don't match, or positive nudges when the codes match a bit better. Then you go away for the night, or the day, or the month, or the year, and next time you look the neural net has generated the right weights so as always to print out the letter of the alphabet that corresponds to the letter it 'sees' in the TV camera. And it seems to have learned to do it all by itself!

Nowadays the procedure has been continually refined by the application of clever mathematics and the right number of layers of neurons in the net to give the optimum kinds of nudges, feedbacks and memories, and it works well for all kinds of patterns with more and better connections and weights being generated by computers engaged in learning to recognize more and more features of the world. With well-designed feedback and learning algorithms, we get a virtuous circle and the computers become increasingly clever. Furthermore, they get cleverer and cleverer faster and faster because of Extended Moore's Law – the rate of change of the number of nodes and speed of training are going beyond our linear imaginations. This exponential increase in speed means that if, in the 1960s, we had to walk away for a year to allow the machine to teach itself to be an accurate recognizer of capital letters, fifty years later we could walk away for only a second or so. And that makes another kind of difference, similar to the way plants appear intelligent if speeded-up time-lapse photography shows them competing with each other as they reach for the sun. For a machine that takes a year to learn recognition of capital letters will still look like a machine, but if it learns in a matter of seconds it will appear intelligent because its speed of learning seems to have surpassed that of humans.

Notice that an essential feature of the system, and this is crucial to the argument, is that someone or something has to tell the neural net when its performance is getting better and when it is getting worse. Consider the very simple

letter-recognizing program we have imagined. So far, we
have thought only about capital letters being presented. But
imagine the device is presented with a lower-case 'a' and
responds with 'A'. Should it be rewarded or punished? If
it is being trained to recognize capitals, the answer is punished; if it is being trained to recognize letters irrespective of
case, the answer is rewarded. So it looks as though a decision
about what we are trying to teach it cannot be avoided – such
devices, it seems, cannot just be built and let loose on the
world; the choices about what to reinforce are a primitive
form of 'enculturation'. But this situation is similar to that
of a child, which also cannot be just let loose on the world;
without enculturation a child's abilities would not develop
very far either. Here, we are drawn back to the question of
where intelligent computers get their understanding from –
just as we have to ask it for humans. Computers, as we said
in the introductory chapter, can be 'brought up' in different
ways.

Pattern recognition: Bottom-up, top-down and the sociology of knowledge

These chapters unite the position of the sociologist and the
AI believer in ways that, a few years ago, seemed impossible.
The ideas we have bounced off are, once more, those of Ray
Kurzweil. In Chapter 7 we'll look at Kurzweil's claim that
the crucial feature of the brain is a huge set – 300 million –
of hierarchically arranged pattern recognizers. I am told by
those who understand neuroscience that Kurzweil's model
is far too simple and there are many other crucial features of
the brain but, for the moment, it is useful to work from the
Kurzweil model even if it is only a 'cartoon' of the brain. In
the spirit of the principles outlined earlier, we will make it

harder for ourselves by assuming Kurzweil's simple model is right. What we want to show is that *even if* this simple model is right, and it is a model that, from the outset, makes the brain into a kind of computer very similar to the computers currently being built, it still won't learn to do what humans do without some kind of modification that enables it to be embedded in society.

Elements of pattern recognition

The specially attractive thing about Kurzweil's model in respect of this book is that a central topic of the sociology of knowledge is also pattern recognition; if brought up in one society one sees one set of patterns in the world – perhaps gods, witches, ghosts, magic – but if brought up in another society one sees mortgages, humans evolving and bathrooms in need of remodelling. The central topic of the sociology of *scientific* knowledge is the recognition of patterns within science; it is the study of how scientists discover and establish new patterns in the world. For example, I have just published a book describing the first detection of gravitational waves. That first detection took between a hundred years and five months, depending on how you look at it. But, even taking only the shorter period, it was five months of debate about whether some vestigial numbers mixed up with noise emerging from a hugely complex and delicate apparatus were to be counted as the first observation of a new kind of pattern – the gravitational waves emitted by the inspiral and merger of a pair of black holes.

Once upon a time scientific pattern recognition seemed simple: brilliant scientific minds invented theories to describe the patterns in the world, and observers and experimenters then confirmed or disproved these theories by seeing or failing to see the patterns; each new discovery and pattern built on and added to the overall pattern. Scientific knowledge was *monotonic*, a term we'll come back to later, and cumulative. The patterns were things like descriptions of the movements of the planets or patterns in the relationship of time, space and the speed of light. Maybe the data came first and the brilliant scientific minds extracted the patterns from it, or maybe it was the other way round. This seemed so straightforward that in the early days of artificial intelligence it was believed that a program called 'the general problem solver' could do science all by itself. A particular program called BACON was said to have worked out Kepler's Laws of planetary motion from astronomical data. In a sense, it had, but, as we'll see, it was a misleading sense.

From around the early 1970s onwards this simple picture of science began to be questioned. Close examination of the work that scientists actually did, day in and day out, showed that the skills involved in doing successful experiments were mostly tacit and couldn't be described formally. In this they were like the skills we use to ride a bike, or pick up a cup, follow a cooking recipe or produce a well-formed sentence in our native language. The same applies to the skill of observation: when you look down the microscope at the pondwater, it takes skill to see that confused mess as a pattern of microorganisms. This meant that no one could be completely sure whether data or observations were real or artefacts: was the experimenter or observer seeing genuine patterns or the equivalent of 'pictures in the clouds'? Furthermore, there was opportunity for different groups of scientists to see different patterns in the same data depending on which sets of experiments or observations they believed had been done competently. The simple monotonic cumulative picture was replaced by a *modulated model*: the modulation was a matter of varying social agreement among scientists. There is a huge literature on this transformation of our understanding of the nature of science, but one can immediately see that it implied that a crucial feature of science was not included in the simple picture of theory being confronted by data; the crucial part was deciding whether the data was real. Close examinations of science showed that scientists in difficult areas spend much of their time arguing about whether some observation or experimental result was genuine, with the scientists often polarized and arguments lasting decades. To give just one example, chosen because the typical incorrect version of things can be found in Kurzweil's 2012 book, the famous Michelson-Morley experiment of 1887 was said to have shown that the speed of light was a constant and had given rise to a puzzle resolved by the theory of relativity. Actually, the experiment was never completed; a decisive experiment of Michelson-Morley kind was not carried out until at least the middle of the twentieth century, the result being much disputed up to at least the 1930s, and the 1887 result had no influence on Einstein's formulation of relativity anyway.[84] In the same way, though BACON appeared to have deduced Kepler's Laws from data describing the movement of the planets, it turned out to be data that had already been through a sieving process informed by Kepler's Laws. What looked like raw data was data made exact and noise-free by having Kepler's Laws already built in, whereas real science is all about separating the observations that have to be taken seriously from those that should be discarded. BACON was just going through an arithmetical procedure in the same way as, say, a factor-analysis program. It was not an artificial scientist at all – real scientists have to find ways to separate signal from noise and that usually involves disputed judgements about *who* to trust as a proxy for *what* to trust![85] We'll put some observational flesh on the bones of this in Chapter 8.

More on bottom-up and top-down

There are two basic models of pattern recognition, bottom-up and top-down, that relate to these competing accounts of how science is done.[86] If you take BACON to have worked in the advertised way, it was deducing Kepler's Laws from data that was found in the world without human intervention (beyond providing the means to allow the data to be collected). This is deducing Kepler's Laws by *bottom-up* pattern recognition. Bottom-up means that patterns are already in the world and they will rise to the surface of our understanding like a heated liquid rising to the surface of an immiscible cold liquid: think of the coloured blobs of oil in a 'lava lamp' rising to the top. The bottom-up model is what informs the simple model of science.

If, on the other hand, the data used by BACON has been filtered by humans in such a way as to remove all the data points that do not fit Kepler's Laws and to retain only those data points that do, the deductions already have a heavy component of *top-down* pattern recognition. For the sociologist of scientific knowledge, science is mostly about top-down pattern recognition – it is about what humans put into the patterns. The sociologist's *ideology* is all about top-down pattern recognition, whereas the *ideology* of the simple model of science – and it is the ideology of at least a subset of artificial-intelligence believers – is that science and observations are all, or mostly, bottom-up. Here we are going to fit those two ideologies together to make a modified model.

Incidentally, for those who want to preserve a simple model of science, the complexities that stand between theory and data uncovered by the sociology of scientific knowledge can be said to show only that the surfacing of the oil in the lava lamp is sometimes very slow – the lamp works very sluggishly. This is the common-sense view of the world and one given enormous force by our seeming ability to control the world around us by finding patterns in it. Notice that it is also the model that underlies the idea of a purely objective social science – a sociology where we could induce the way of being of an alien society on a poisonous planet, or the way of being of a strange terrestrial society, solely from outside observation without any messy 'understanding the natives' or 'subjective' interpretation.

Now, the existence of all this top-down messiness in what was once thought to be the socially isolated domain of science has been discovered through the close examination of the way science works. But you don't have to believe it. The sociology of scientific knowledge took science as the 'hard case' – where if these social processes could be shown to apply, then they certainly applied everywhere else. But even if you are not prepared to accept even a hot, rapid, lava-lamp model, where the oil reaches the surface quickly and with almost no influence from society – in a deterministic, near calculative, way – and you are not interested in the smallest amount of social stuff, you still have to accept what is being said in respect of all the forms of human knowledge that are not science. For art, for fashion, for literature, for witches and gods and parapsychology and alternative medicine, it is predominantly top-down. Actually, you even have to accept it for science because hardly any science is represented by the Newtonian-physics model that gives us Kepler's Laws and the rest, nor by the Einsteinian or quantum theory examples. Most science is full of uncertainties – weather forecasting and econometric modelling would be better paradigms for science in general.

To reiterate, the top-down model underlies interpretative sociology – the kind of sociology which begins with understanding how people live their lives in different societies. The model says that patterns are imposed on the world and those patterns will vary from society to society. What counts as a set of the same (polimorphic) actions in one society or social sub-group will not count as a set of the same actions in another. Here saying 'you bastard' will count as a greeting; there it will not. Only mimeomorphic actions, where the same action is always executed with the same externally visible behaviour, can give rise to generalization and predictions from the outside. The ideology of the interpretivist holds that there aren't any patterns 'out there' and the patterns that we perceive are all imposed on supposed reality by us – it's *all* 'pictures in the clouds'. This view can be retrieved any time it starts to seem tenuous just by thinking of the groups for whom witches, ghosts and gods have as much reality in their lives as anything else. The same goes for judgements about beauty or goodness. More generally, the view is hard to resist when we remember that in polimorphic actions we group things – actions (this action is the same as that action) – according to their social meaning, not their physical substance or spatial and temporal order. Social meaning, of course, varies from society to society, from group to group and from occasion to occasion. The problem is an ancient one going back at least as far as Plato, with a solution still being sought by the later Wittgenstein – what is a table or a game? If a table is described by its bottom-up characteristics – flat surface supported by legs – what should we call the tree stump that serves for a picnic? As for the meaning of 'game', Wittgenstein solved the problem by suggesting we look to the use, not the meaning; uses pertain to societies.

The sociology of scientific knowledge (SSK) took top-down pattern recognition (otherwise known as relativism or social constructivism), as a credo which powered the battle with the preceding simple view of science. SSK claimed that humans created patterns, including scientific generalizations, through their socially agreed choices about what was to count as sound observation and what as unsound. I am going to

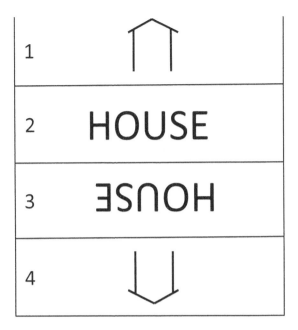

Figure 6.1 *Four ways to represent a house*

represent this idea in a modified form that fits what we have to accomplish here.[87] The argument starts with four ways of representing a house, as shown in Figure 6.1.

Staring with the first icon, the sociologist would be inclined to say that it looks nothing like a house – take it into the street and compare it with a house! The determined 'bottom-upper' will insist that there are similarities between the house and icon 1 – such as that between the slope of a house roof and the upper lines of the icon. In response, the determined sociologist could point to the second way of representing a house. There really is no resemblance between the way of representing a house in the second line of the figure and a real house – the relationship is all top-down. Just to show how true this is, we can imagine that if our social conventions had been different and our language had developed in a different way, a house might have been represented by MAISON or CAT, or by the symbols shown in the third line of the figure. If symbolic representations of houses can work like that, why should icon 1 not be equally based on social convention alone?

But now we run into a difficulty. I can recognize what is in line 3 as the same symbols as in line 2 but inverted. Yet, before writing this account I had never seen the word HOUSE written upside-down, so the relationship between lines 2 and 3 (forget the relationship to actual houses for the moment) cannot be all top-down. That I can see the similarity of the symbols in line 2 and line 3 must involve some bottom-up recognition. There is something 'out there' in the relationship between line 3 and line 2, or I would not be able to see the similarity that allows me to say they are the same apart from one being on its head. Without something being fixed in the world, those two sets of identical 'symbols', one inverted, would simply be two unrelated patterns.

If there is something 'out there' in the relationship between lines 2 and 3 it opens up the possibility, once more, that there is something 'out there' in the relationship between the house icon and a real house. If that is true, there will almost certainly be less relationship between the icon in line 4 – which is icon 1 inverted – and a real house than icon 1 and a real house because real houses are the right way up.

In that case, icon 4 would have had a harder time becoming used as a house icon than icon 1. (But the symbols in line 3 would not have had a harder time than those in line 2 if we imagine that the entire alphabet had developed in 'upside-down' form.) We can describe the extra stress and strain which would have attended icon 4 becoming accepted in society as the house icon as its lack of 'affordance' when compared with the first icon. Icon 1 'affords' house better than icon 4 and, of course, better than lines 2 and 3, though, in our society, in the case of line 2 (but not line 3), this initial lack of affordance has long been overcome through familiarity.[88]

In a published discussion with Arthur Reber, a psychologist whose thinking is very much informed by the idea of evolution and who is an expert on the psychology of tacit knowledge, I found myself having to agree that, in spite of my sociological *ideology*, to deal with any kind of bottom-up pattern recognition – such as is needed to explain why line 3 is line 2 inverted, or why icon 1 is a better icon for a house than icon 4 (assuming it is) – I would be forced to adopt something like Reber's evolutionary starting point.[89] I would have to agree that animals, as they emerged from the slime, evolved to extract certain basic shapes from the environment (such as the upright triangle representing the roof in the house icon). That is the only way I could explain, say, the greater affordance of icon 1 as compared to icon 4, and the immediately evident relationship between lines 2 and 3. Adopting Reber's approach, I would predict that upright triangles, since they must have become recognizable as distinctive entities so early on in human evolution, are likely to be recognized across all cultures. There must be, I would have to argue, a substrate of fundamental shapes, patterns and perhaps colours and textures that all humans recognize and upon which the variations in culture are built. This kind of recognition of patterns would not be subject to the sociology of knowledge because it would happen across all cultures – it would be bottom-up!

Reber told me that Biederman, in 1987, developed a model of perception based on primitive forms like circles, cones and ellipses called 'geons'; real houses and house icons (and presumably letters of the alphabet) can be seen as composed of such forms. Biederman's theory uses less than forty geons to account for virtually all relatively fixed objects. We are not endorsing the geon-model in particular, but taking it as a paradigm; there are similar schemes that resemble it for which we will use the label 'geon-like models' as a general description. As we will see, it is very close to some of the theories developed by Kurzweil and his colleagues. Once more: a characteristic of geon-like models of pattern recognition is that they are universal – the same patterns will be recognized irrespective of the culture in which the recognizing is being done.

Most of our perceptions and conceptions still take place within cultures, and the patterns they support are generated from the top-down, with line 2 in Figure 6.1, the upright *word* for house, as the paradigm. But we now recognize that there could be no top-down pattern recognition without it resting on a bottom-up geon-like model which is the necessary condition for any 'symbol', or icon, to be stable and recognizable in the first place; if you don't have recognizable, stable symbols you cannot develop cultural conventions that give them meaning.

It is important to understand what role a *necessary condition* plays in an explanation. Human perception depends on many necessary conditions but none of them do much when it comes to providing explanation of human perception. For example, it is a necessary condition of human perception that blood circulates in our bodies to keep our sense organs alive, but we never find ourselves citing blood-circulation when we talk about how humans perceive the world – we just keep this necessary condition in the unremarked background. We need to think about whether bottom-up pattern recognition is a really important part of the explanation of perception or just a background condition like blood-circulation.

The distinction between bottom-up and top-down pattern recognition will be important as we navigate our way through contemporary developments in AI. Because of the importance of top-down pattern recognition it is vital to embed AIs into social life if they are to acquire human-level fluency.

Nevertheless, we must have some bottom-up recognition to make sense of how any perception works. I am going to call the top-down model, with some bottom-up necessity, the 'modified sociological ideology'. It is 'modified' because the determined sociological ideology concentrates on top-down alone.

Geoffrey Hinton's pure bottom-up perspective

But could all pattern recognition be done from the bottom-up? After all, the creators of BACON thought they had built a machine that had executed a purely bottom-up derivation of Kepler's Laws. Computer intelligence is a good way of thinking about the question. How might one make the case? 'Deep learning' is pattern recognition by machines based on an enormous number of examples of images limited only by computational resources (which, we are assuming, will soon cease to be a problem). It is claimed that computers can become very reliable at recognizing future instances of patterns, such as pictures of objects seen in photographs, when they have been trained on huge databases. It is said that deep learning enables computers to recognize new instances of the same things from new data sets with 95 per cent accuracy or more.[90] It is also said, at least by Geoffrey Hinton, a founding father of the field, that pattern recognition is deeply rooted in the world and that the computers can do it *without supervision* – '*unsupervised*'; Hinton thinks it is all bottom-up.[91]

By the way, Hinton's definition of 'unsupervised' is a much stronger one than is generally used in the deep-learning community. There, the term 'unsupervised learning' includes anything that is not 'supervised learning', and 'supervised learning' implies deliberate intervention in the training of the computer – it means putting deliberate and self-conscious work into the training, though there is a large spectrum for how much work and of what type might go into it. I will introduce the term 'implicit supervision' for all supervision that is not self-conscious and deliberate but creeps in as a result of the way humans, in their ordinary lives, arrange the materials that computers encounter.[92] Most of what deep-learning computers do is certainly not completely unsupervised bottom-up learning as in Hinton's definition. Nearly all of it is either implicitly or explicitly supervised, the latter being described as follows by a computer scientist:

> ... thousands of people are paid pennies to create a 'ground truth' by providing labels for large data sets of training examples. . . . In this case, the 'objective function' is no more or less than a comparison of the trained model to previous answers given by the [humans]. If the artificially intelligent computer appears to have duplicated human performance, in the terms anticipated by the celebrated Turing Test, the reason for this achievement is quite plain – the performance appears human because it is human! . . . The artificial intelligence industry is a subjectivity factory, appropriating human judgments, replaying them through machines, and then claiming epistemological authority by calling it logically 'objective'.[93]

Going back to maximally unsupervised learning, Hinton was quite explicit about his view being opposed to what he referred to as 'the strong Whorfian hypothesis'. This is a reference to what is often called the 'Sapir-Whorf hypothesis', which is closely related to what we have been calling 'top-down pattern recognition'. The standard example is the Inuit who are said to have seventeen different words for snow because of their cultural propensities, whereas we have only one or two. The empirical basis for this has been questioned but the principle is clear – different societies recognize patterns in different ways. Hinton is insisting, instead, that the patterns are just 'out there' before there is any interaction with human cultures and that is why computers should be able to recognize them, on their own, *without supervision*. Hinton said: 'It completely destroys what psychology [I would have said sociology] has said for years . . . [the] categories are really there.' Hinton was also very concerned to oppose 'relativists' who think knowledge is culturally specific; bottom-up learning is universal not culturally specific.

As I will explain, I cannot make sense of Hinton's claim that there is no need for top-down recognition. The view is certainly interesting, however, and worth looking at in some detail because it takes us into the heart of the debate. We now know from the argument about how we represent houses, that we need some bottom-up processing if we are to have a valid model of how the world works; the questions are: how much can be done by bottom-up processing alone, how much top-down processing do we need, and how much pattern formation has been inserted in a hidden way in the form of implicit supervision as humans build the deep-learning algorithms?

One of the most interesting and challenging things pointed out by Hinton was that even if deep-learning computers are presented with objects created by humans, they can still separate them into categories by themselves. His example involved the digits 0–9. He said that if many examples of these digits were handwritten by different people and presented to the right kind of neural net analyser, without any kind of prior training or any supervision it could sort them into sets corresponding to the natural numbers: all the ones would be in one group, all the twos in another group and so on, based purely on presentation of images comprising matrixes of pixels. It would recognize the distinctiveness of the numbers, one from another, because that distinctiveness

was 'out there'. (The same would go for patterns of sound –
phonemes – he said, and that is how speech recognizers could
work.) Note, then, that even if the numbers and so forth
were the products of human culture, the ability of machines
to recognize them as distinct patterns would be universal!
A machine previously exposed only to non-literate cultures
would still be able to separate the numbers one from another
because this separation does not depend on any top-down
guidance; to the extent that number recognition is sociology
in action, we could accomplish that small part of the sociology of alien societies from the outside, just so long as we can
recognize what counts for them as symbols.

Presumably, Hinton believes that humans construct their
written symbols as they do because they too rely on bottom-
up differences in the world. That seems entirely plausible
under the modified sociological ideology which recognizes
that some kind of geon-like model is necessary: when humans
invented the symbols they use in writing they chose patterns
that would be readily distinct, given their biological propensities to recognize basic geon-like shapes. This is what allows
lines 2 and 3 of Figure 6.1 to have a readily recognizable
relationship.

There seems to be a problem, however. One can imagine
such a system distinguishing the numbers based on no more
than pixel intensity since there would be white spaces between
the symbols. But there are societies where a single digit is
represented by two symbols side by side, with white space in
between, such as – I am inventing this example – '00' standing
for zero instead of a simple '0' and so on. In that case it is hard
to see how the machine working by itself would know this was
one number rather than two. It may be hard to imagine, but
one must always be ready for an ingenious solution and in this
case the constant co-location of two zeros might do the trick!

But even setting that possibility aside, it seems that before we can go much further we have to introduce an element of top-down recognition. An unsupervised computer – instructed only to separate patterns of pixels into different groups – will not know that the groups into which it has sorted the number-images are anything to do with numbers. Thus, if we include in the data-set versions of 'A–Z', 'µ', '£', or anything else you can find when you tell your word-processor to 'insert symbol', we have to assume it will sort those into separate categories just as it sorted the digits. But, at first sight, it will have no means of separating the set of numbers from the letters or any of the other sets of symbols. There will be no next level up in the hierarchy of pattern recognizers – that is, memories based on experience (such as is found in the Kurzweil model of the brain, described in Chapter 7) – to separate numbers from letters and other kinds of symbol. At the next level up – for example, numbers versus letters – all the symbols will seem the same. Initially, it seems that to sort into categories at the next level up will require top-down input in one form or another. That is why I was not able to make sense of Hinton's explanation of how things work.

But let us do our best to imagine an answer. It might be said that the equivalent of top-down input can be garnered from the internet by some bottom-up statistical procedure. We now know that we must always expect that some ingenious brute-strength method might be found to resolve a problem at Level I of artificial intelligence – engineering – or Level II, an asymmetric prosthesis. Thus, a program might recognize the way the different kinds of symbols are grouped within documents. For example, strings of numbers without spaces tend to be haphazardly ordered and of widely differing lengths, whereas strings of letters tend to be of limited length and exhibit regularly repeated patterns, while 'µ' is found predominantly in documents alongside lots of other symbols from the limited set of what we call Greek letters; '£', along with '$', and quite a few others are generally found singly at the front end, and only at the front end, of a string of numbers. This is the kind of precedent-based statistical procedure that usually takes people inhabiting the right-hand side of the expertises represented in Table 2.2 (p. 30) by surprise because we are less good at foreseeing such things than are the experts on the left-hand side; here I am *trying* to do left-hand-side thinking. This higher level of classification that we have invented would be universal – you do not have to know about numbers or possess the *idea* of Greek letters, or of any other kind of symbol, to do this type of statistically based sorting, so it could be done without reference to the cultures that use numbers and these other symbols. The sorting could be accomplished purely by recognizing strings of symbols within a body of printed documents and it could be done with no other supervision than being 'rewarded' for separating symbols into types by finding patterns in their use within the documents.

But this is where it gets tricky again because this kind of sorting is not entirely bottom-up,, even though it could be accomplished 'without supervision'. The humans who created the documents in the first place created the orderings and associations of symbol-use found in the documents by reference to the cultures within which *they* were socialized. This is analogous to BACON finding patterns in data that had *already been sorted* by reference to Kepler's Laws. In other words, to know the different meanings of various kinds of symbol, even if by 'meanings' we intend no more than knowing how to categorize them into sets – numbers vs letters etc. – we need to have seen them *in use* within human societies. This, once more, is implicit supervision.

The categorization produced by this means will not, after all, be universal in the way the recognition of the individual symbols was universal and that Kepler's Laws are (putatively) universal, because it will vary from culture to culture. Thus, imagine there is a culture that uses the symbols A to I in place of 1 to 9 and vice versa, leaving everything else the same as our convention. In that culture the categorization will be different: some of what we call numbers will be classed with letters and vice versa. And then there are languages/cultures such as Hebrew and Roman, where the same symbols are used as digits and numbers, so sorting them is always context-specific. For unsupervised computers to extract culturally embedded human categorizations from documents is implicit supervision because it depends on features of the culture being pre-inserted into the database. It means deep learning is relativistic (in the slightly weakened form consistent with the modified sociological ideology). Implicit supervision is top-down and relativistic, even though it might not imme-diately appear to be so. This is an example of the entry of top-down pattern recognition in subtle and hidden ways for which we must be continually vigilant if we are to understand the world of the new artificial intelligence. But, given this, sociologists have to admit that more than they thought can be learned about alien societies purely by studying them from outside. We always knew we could work out the demography of an alien society by counting its members and maybe still more by watching their movements, but now we know we could also learn something about how they use symbols – not much more, but more than we thought!

To repeat, under the 'modified ideology' some bottom-up recognition is going to be necessary to allow the individual symbols to be recognized in the first place, and it is *sufficient* to allow the individual symbols to be recognized in the first place, but it is *not* sufficient to separate the symbols into types without top-down guidance, however subtle. If com-puters can learn from the human world with only implicit supervision it is still a hugely important accomplishment but it makes what computers learn parasitical upon human societies and consistent with the Sapir-Whorf hypothesis. At least, this is true for the computers we have, forgetting about those that might develop independent societies with completely independent and incomprehensible realms of knowledge. For the computers we have, it is vital that we understand the top-down input if we are to comprehend what is going on in the world of intelligent machines.

Hinton and I also discussed the recognition of other kinds of pattern. For example, Hinton insisted that deep-learning programs could recognize and classify women separately from men – the example that happened to come up in the conversation. Here is a transcript of the conversation:

> *Collins:* what happens if you train your device to recognize all women and then show it an icon of a woman such as you might find on a lavatory door?
>
> *Hinton:* Are there icons in the training?
>
> *Collins:* Let us say 'yes'.
>
> *Hinton:* Then it will be fine, it will include those with the women.
>
> *Collins:* How will it have worked out by itself that the icon of a woman is to be classed with a woman?
>
> *Hinton:* Is it supervised or unsupervised?
>
> *Collins:* Unsupervised – it is the unsupervised bit that is getting me [i.e. that I am failing to understand – I say this because insofar as there is supervision then top-down pattern recognition is finding its way in via the system of effective rewards and punishment provided by the supervision].
>
> *Hinton:* What it ought to do is represent the icon as having several properties, one of which is that it's a woman and the other of which is that it's an icon. So all the good

unsupervised algorithms don't just classify – right – they describe. So they get much richer representations than just . . . they're not just clustering. Most primitive unsupervised learning is to cluster things. But that's a very weak form of unsupervised learning. The only description you then have . . . is of which cluster it belongs to. That's not much information. The good unsupervised learning algorithms produce . . . a description. What you want it to do is to say 'this is a person, it's a female person, and it's an icon not a real one'. It will produce a big vector of features that will cover all those aspects.

Collins: But you're saying that all by itself, without any supervision, it is going to be able to associate icons of women with women.

Hinton: Yes – they've got a lot of structure in common.

Collins: I can't get my head round it – what sort of structure would an icon of a woman have in common with a woman?

Hinton: Well, let's make it a bit simpler to begin with, right? Let's say 'what sort of a structure would a sketch of a face have in common with a real face? . . . an ellipse with two circles, the hair and the mouth?' Take that: what does that have in common with a real face? Well, the answer is it has an awful lot in common. It's got the same arrangement of elements, and the elements themselves are roughly the same shape.

So Hinton is saying that a good learning algorithm will pull out features of what it is recognizing, and that the icon of the woman and actual women have enough features in common to allow such a program to recognize them as belonging to the same set of objects in the world. The same would apply, of course, to houses and their icons – there would be enough in common between a house and icon 1 in Figure 6.1 to allow an unsupervised program to recognize the icon as belonging in the same set as images of houses. When I write on p. 113, 'This icon looks nothing like a house – take it into the street and compare it with a house and it will be obvious', I must be wrong if Hinton is right. According to Hinton, the icon looks enough like a house – the slope of the upper lines, for example – for a computer to be able to group it with other houses and, given the need for geon-like models, we do not have to object to this possibility in principle.

And yet even if the real women could be classed with their icons by a deep-learning program in a bottom-up way, and even if the same goes for real houses and their icons – even if we accept all this – the argument, once more, does not go very far. We know this because 'HOUSE' has no bottom-up features in common with a house (it could be 'CAT' and in France it is 'MAISON'). Furthermore, in Japan, the icon that is sometimes used to represent a man on lavatory doors is a smoking pipe, and there is nothing in common between a smoking pipe and a man that can be recognized solely from the bottom-up. Or maybe there is: maybe a computer with access to the internet could find that in Japan there are lots of male images associated with smoking pipes, but again this is implicit supervision not bottom-up pattern recognition, because it would depend on the humans writing the documents having been guided by their culture.

But let us remind ourselves of what can be achieved by statistical, precedent-based methods. It has been found that documents can be usefully analysed if they are thought of simply as 'a bag of words'. With this knowledge, and without any attempt to make sense of those words, it is apparently possible to separate email 'spam' from useful emails ('ham'). So imagine our pattern recognizer is given a body of emails and told only to separate them into classes based on the statistics of word usage with the documents. Presumably such a program would separate the emails on many dimensions – long words versus short words, number of foreign words, proportion of adjectives and so on – but one of those dimensions would be spam vs ham, recognized by sets of words that are more usually found in spam than in ham. That seems like purely bottom-up classification but, as with the difference between numbers and letters, the program does not know either that there is a spam vs ham dimension, nor which is which. To pull out and dump all the spam emails from that set of different classifications requires that some human 'tells' the program, basing the instruction on that human's understanding of the culture of email, that one of the groups of emails it has separated out is the one to be dumped – thereafter, but only thereafter, the program should be able to do it by itself.

Deep-learning programs also learn to recognize images of cars, animals and other such things. That is, once trained on a huge set of images, it is said the programs can recognize new instances at levels of accuracy in the 90 per cent plus range. We have no reason to disbelieve this. But what we don't know is how much implicit supervision is going on in the training sessions. Are the separate objects in the images labelled by humans (they usually are) or is the computer picking out separate objects by itself in the way it could separate symbols into discrete groups? If the objects are labelled, then there is supervision going on. This is true even if the pattern recognition is based on a 'bag of images' model. This is like the 'bag of words' model but uses fragments of images, a bit like geons but without the geometrical purity – they might be a typical bit of zebra where the back leg joins the body, a bit of a motorbike, where the front axle enters the wheel, or a bit of a violin such as the bridge with a few strings and a section of the curved sound box. We could imagine that the computer can pull apart zebras, motorbikes and violins based on the statistics of the image-elements found in different images, even though it would not 'know' which groups were, respectively, zebras, motorbikes and violins without being told, or having them labelled at the outset – just as with spam and ham.

But even the first step is not quite so clear on closer examination. Going back to the sorting of emails, it is assumed that the program knows where each email begins and ends: it must either be told this or told to work it out for itself – something that we can imagine it doing given that emails have readily recognizable headings and terminations. But images do not have such obvious termination points. It is possible that all the work done on deep learning of image recognition has depended on training with sets of photographs found on the web, where the image is where it normally is when we take a photograph, centred or near-centred and complete or largely complete. So the limit of the bag into which the images are loaded is defined by image position in the photograph, and the position of the photograph is a representation of human culture – we take photographs in a way that reflects how we have already sorted the world into different objects according to the way we have been socialized.

Now imagine if, instead of drawing on these human-influenced photographs, the computer was presented with a set of images taken at random. Such a set would be expected to include only bits of zebras and bits of motorbikes and bits of violins, each photograph also containing bits of other things and many such bits overlapping. Presented with such a set of photographs, it is not so clear that the computer could do the original classification – just as if we removed all the identifying features that separated one email from another and gave the computer a continuous string of text, we would not expect it to be able to take even the first step of categorizing them.

I asked Hinton if the deep-learning programs would still work if the images were random shots of the world. He said that the training sets did use centralized images but that it would still work if the images were random, only the training would take longer. This looks like something that needs testing empirically – for example, comparing the time taken and degree of success for a training set that has been used already with one where the images are sliced up at random so that the objects are no longer central or complete as when images first enter our eyes and cameras.

It is, therefore, hard to know if this suggestion regarding an implicit top-down categorization in the images that are being sorted is correct. But it illustrates how hard it is to separate top-down and bottom-up, a task that is vital if we are going to understand our world.

It seems that the way that bottom-up and top-down pattern recognition work together is a central issue to enable understanding of the new artificial intelligence. We have seen strong claims made for all pattern recognition being bottom-up and that deep learning can work without supervision, but very careful analysis is needed to see if this true. The implicit claim is that computers left to themselves would rediscover the same world as we humans have discovered because the process is the same for both human and non-human intelligence. That is the anti-Whorfian hypothesis.

But what is the thing that it is supposed we would all discover if left to ourselves? Is it the world according to the simple BACON view of science? Is it thought that a computer left to itself would rediscover Kepler's Laws and, by extension, every other scientific fact that humans have discovered? This model of science is the monotonic one – there is only one science and, indeed, there is only one knowledge. This is a curious view given the existence of scientific controversies that last for half a century, and the multiplicity of different knowledges in the world. I suspect that it is believed that the single knowledge that we would all discover is the knowledge gathered on the internet in Wikipedia and other such sources. If we do come to be persuaded that this is the single true knowledge, it will be the most remarkable achievement of Western imperialism yet.[94]

To sum up a point that it has taken me, anyway, decades to grasp:

Humans interpret their elementary perceptions according to the conventions of their culture; the elementary perceptions that humans interpret are universal.

To neglect either side of the proposition is to misunderstand both AI and the way the world works. That this is the way the world works can be checked simply by thinking about Figure 6.1.

We have to be vigilant so as not to fall into the trap of thinking that Western culture is itself the universal thing. If top-down pattern recognition is creeping into deep learning and all the rest we must not mistake it for elementary perception; we need to be able to spot the top-down, even if it is hidden or implicit. But this is not a trivial task – it requires knowledge of exactly how the machines are trained and this is knowledge belonging to the experts on the left-hand side of Table 2.2 (p. 30). Analysts from the right-hand side can only ask the questions and venture some answers, but the real authority has to come from the left-hand side – a left-hand side willing to analyse their own favoured project as half-empty rather than half-full.

7

Kurzweil's Brain and the Sociology of Knowledge

In this short chapter we try to widen the meeting between deep learning and the sociology of knowledge. We start with Kurzweil's proposal that the main feature of the brain is a hierarchically organized set of pattern recognizers; this is set out in his book *How to Create a Mind* (2012). If we were worrying about Level IV of AI we would be concerned about whether Kurzweil's description was right, but in this part of the argument we are looking at Level III. As far as this level is concerned it does not matter too much whether his model is right, or if it is a kind of 'cartoon' brain and the real thing is actually much more complicated; we are going to assume Kurzweil's model is correct because it nicely draws out the principles being examined here. What is being argued here is the need for social embedding – the central argument of the whole book – and nothing would change if the brain is actually much more complicated. We represent Kurzweil's model in Figure 7.1, as a series of levels of neurons with mostly vertical links.

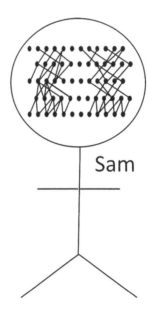

Figure 7.1 *Sam's brain as a hierarchically arranged series of pattern recognizers*

In Figure 7.1 the stick-person represents an individual who, for ease of exposition, we'll refer to as Sam; Sam's head has been drawn very large for ease of presentation. Sam's neocortex is represented as five rows of nodes with some up-and-down links between them. Two clusters of links are shown to indicate that there is some 'sideways' specialization in the brain – not every node is linked to every other.

At the lower level of this model, bottom-up pattern recognition is the starting point and, without being exact about it, we can take it that it is done by the lowest level of neurons facilitated by some geon-like model; Kurzweil himself talks of fundamentally recognizable elements in the world, such as vertical lines, horizontal lines, lines sloping one way or another, curves with their convex faces left, right, up, down and so on. These initial patterns are then assembled into larger patterns by up-and-down feedback to the next higher level. The process is repeated, layer by layer, all the way up to the top level which stores our experience in the form of memories and reinforces or vetoes what comes from below. The first higher level might be letters of the alphabet; a higher level than this might be whole words and the next higher level might be phrases. Each level feeds backwards and forwards to higher and lower levels, and thus the capture of even whole sentences can be understood as a process of hier-archical pattern recognition, and the same applies all the way up from bits of images at a low level to complete images at the top. Memories store our lifetime of experience and they affect the likelihood that some input at the bottom level will be discarded or accepted as representing some high-level pattern that corresponds to that experience. If the input as it passes up the hierarchy continues to trigger positive signals from the higher levels until it reaches the top, and then triggers the response appropriate to a familiar object, say a duck, then we will see a duck. If it cannot trigger anything familiar when it gets to the top level, then it greatly increases the chance that the whole train of signals will be rejected. It is surely much more complicated but those are the principles.

In the language used here, the lower-level pattern rec-ognition in Sam's brain works mostly bottom-up, but the higher level works mostly top-down. As Kurzweil says, it has to depend on *memories* of higher-level patterns stored in the brain. And those higher-level patterns are established accord-ing to the society in which one is brought up, or 'socialized'. One can see how well Kurzweil's model of the brain fits with the modified sociological ideology described above and how it has the potential to fit with the sociology of knowledge.

Now – we can make this a perfect fit simply by adding another layer of 'neurons' on top of Sam's existing brain. This level comprises the neurons in all the other brains with which

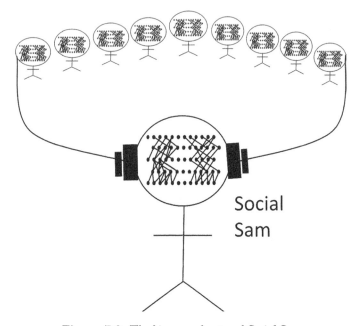

Figure 7.2 *The big neural net and Social Sam*

any one brain is in contact. The modified model, which we'll call 'Social Sam', is represented, much simplified of course, in Figure 7.2.

At the top of the simplified figure are more neocortices representing the vast number of brains belonging to other human beings encountered on Social Sam's journey through life that help to create Social Sam's conceptual world and the components of Social Sam's knowledge: they will be dominated by those in the various social groups that Social Sam inhabits. They will be grouped and interlinked in many ways, some of which will be discussed in the next chapter. That level is, then, Social Sam's society and it is responsible for the top-down influences.

Social Sam's neocortex is linked, backwards and forwards, with the neocortices at the top through the senses of Social

Sam and the others. Like the top layer of neurons within Sam's skull, and Social Sam's skull, the additional level of pattern-recognizing nodes has powerful rights of approval or veto over patterns proposed for recognition by the rest of the nodes. Those rights of approval and veto impact on a new child's brain from birth or before; from outside the skull they shape the experience and memories that form the top level of the Kurzweil brain. I am taking pains to describe all this in exactly the way Kurzweil described the brain, except I have added one additional level at the top which happens to be outside the skull but is nevertheless linked to the neurons within it in one large neural net.

Because the societal level is outside Social Sam's skull the connections between it and Social Sam's brain are sparse and slow compared to the connections within his neocortex, but they are extraordinarily important in the formation of perceptions. To make matters more complicated still, the top level is linked to the rest via the senses and this means its effects must enter initially via the bottom neocortical level. How the top layer comes to have its powerful high-level influence is, therefore, complicated. We can, however, illustrate the logical possibility and power of the process quite simply. If Social Sam is never shown either a cricket bat or a picture of a cricket bat and never hears a cricket bat described, nor reads about cricket bats in the course of his upbringing, there will be no cricket-bat templates in the upper level of his neocortex. This is one crude way in which the upper level outside the skull – all those brains in society – can have a huge and pretty direct influence on the upper layers of the neocortex, even though the influence enters through the bottom layer.

Of course, most of that influence enters via language, mediated in subtle ways, and is most economically described in terms of the establishment of what counts as *sense* in Social Sam's society and in Social Sam's *mind*. But here we are trying to make it hard for ourselves by avoiding notions like 'sense' and 'mind' and simply sticking to pattern recognition. We want to stick with a modified version of the Kurzweilian model of the brain because we want to mount the argument from within the position of those we are trying to convince, and these include experts on the left-hand side of Table 2.2 (p. 30); that's why we don't want to use terms like 'sense' at this point, but stick with electrical or chemical connections. Social Sam as an individual is made up of the influence of many sets of external neocortices; some big, like the one from which natural language is drawn; some medium, such as sport or recreational groups; some small, like the ones from which esoteric expertises are drawn.[95] We can imagine all these external neocortices linked into his brain in the way the one shown in Figure 7.2 is linked. We are still talking of nothing more complicated or metaphysical than neural nets and pattern recognition: it is just that to understand what happens in Sam's brain we need to extend the neural net beyond the brain. To understand Sam, we need to look at the *big neural net* in which it is embedded – we need to understand *Social* Sam. That is the central claim of this book and Figure 7.2 is its iconic representation.

We know that there is a lot of suspicion about the notion of 'society', but there is nothing vague or metaphysical about what is being said here. Nor is there anything 'socialistic' that could threaten the freedom, power and potential success of the individual or individualistic political systems. Or, at least, if there is something threatening, that's just the way things are. This other level is just there in the physical description of the world and its brains whether one likes it or not. That is why people who live in one place and one society perceive quite different worlds from those who live in another: there are mortgages here, witches there; souls in cowrie shells here, a determination to remodel the bathroom and ocean-going yachts that demonstrate God's favour there; gravitational waves that have been detected here; and false claims about the supposed detection of gravitational waves there.[96] That is why there can be a sociology of knowledge and even a sociology of scientific knowledge. All this seems completely obvious and makes a determination to disprove 'the strong Whorfian hypothesis' seem still more strange. Hinton is right that one cannot make sense of a world in which there is no bottom-up influence, but one cannot describe the world we live in without a huge element of top-down influence too. They have to work together if we are interested in describing the world correctly. If one starts with a model of the brain as a hierarchy of pattern recognizers, to describe the world correctly one has to add that extra layer. Describing the world correctly is surely a necessary starting point if we are to understand the actuality and potential of artificial intelligence.

What I have argued here turns out not to be new. Yoshua Bengio points out to me that in 2012 he wrote a paper entitled 'Deep Learning and Cultural Evolution', which argued that connection into the culture was necessary for understanding language.

> We propose a theory and its first experimental tests, relating difficulty of learning in deep architectures to culture and language. The theory is articulated around the following hypotheses: learning in an individual human brain is hampered by the presence of effective local minima, particularly when it comes to learning higher-level abstractions, which are represented by the composition of many levels of representation, i.e., by deep architectures; a human brain can learn such high-level abstractions if guided by the signals produced by other humans, which act as hints for intermediate and higher-level abstractions; language and the recombination and optimization of mental concepts provide

an efficient evolutionary recombination operator for this purpose. The theory is grounded in experimental observations of the difficulties of training deep artificial neural networks and an empirical test of the hypothesis regarding the need for guidance of intermediate concepts is demonstrated. This is done through a learning task on which all the tested machine learning algorithms failed, unless provided with hints about intermediate-level abstractions. (Bengio 2012; <https://arxiv.org/abs/1203.2990>)

It is gratifying that the completely independent approach of deep learning and evolutionary algorithms has also given rise to an understanding of the need for the notion of culture if human knowledge is to be captured.[97]

It seems to me that the discussion of what I am calling 'the big neural net' leads to a choice: we can think that AI is trying to produce the equivalent of an individual brain, as in Figure 7.1 – in which case it will fail because brains are not isolated; most of what they know comes from being embedded in cultures. Much more novel and interesting is to see deep-learning neural nets as trying to reproduce a whole society of brains: something like Figure 7.2 but more so. It is 'more so' because even when one realizes that individuals are plugged into society, the whole of society is much more than the top level of even Social Sam's neurons as represented in Figure 7.1 because even Social Sam doesn't draw on all the brains in society except, arguably, in the case of something like his fluency in natural language where every member of society is continually contributing to what the language is becoming. Mostly, humans are plugged into various small subsets of society's brains, even a single society's brains – it depends on what sub-groups they belong to. So that presents a problem for neural nets if they are thought of as encapsulating all the knowledge there is in a society. If human knowledge is going to be reproduced, some way will have to be found of dividing up all the knowledge there is into the subsets, which appear to be necessary if human society is to function. Subsets of human society work by excluding other subsets – and they have to. For example, if, in science, various subsets were not continually excluded we would have no science because the cacophony of competing views would drown out progress. Human society has, continually, to solve the problem of who to take seriously.[98] The next chapter looks at how working in groups is managed among gravitational-wave scientists. I chose these scientists because I know them well; they can be used as a case study in respect of most frontier science.

8

How Humans Learn What Computers Can't

It has been argued that it is a mistake to think that computers, even computers with human-like bodies, could re-create our world, solely via bottom-up recognition of the ready-made patterns found within it. Humans themselves extract varying patterns from the world according to the multiplicity of cultures found across human societies and there are different cultures even within societies. The same is even more clear given the violent disagreements between scientists from the same Western culture analysing 'the same' data in the same kind of controlled and circumscribed way. There exists an enormous number of interpretations of the world and an enormous amount of 'interpretative flexibility' within any broad interpretation. In any case, the great majority of thinkers about these matters is ready to accept a good component of top-down pattern recognition, and this includes most of those who are sure that deep learning, or something like it, can eventually reproduce human intelligence.

A human learns from human culture

Throughout the book it has been pointed out that insofar as it turns on social embedding, the question of artificial intelligence, which we have argued is the central problem, is the same as the question that faces sociology. A sociologist, assuming that bottom-up pattern recognition cannot work by itself, has to acquire the patterns that inform a society from the top-down if that society is to be sociologically understood. To develop a thorough understanding of a strange society, at least the top level of the 300 million pattern recognizers in the investigating sociologist's brain has to accumulate similar memories to the brains of members of that society, and where those patterns of acceptable usage are changing, it has to keep up with those changes in real time.

This is also a problem that faces the citizen. The Periodic Table of Expertises (PTE) attempts to create an exhaustive classification of all the kinds of expertises that people can have.[99] Included are 'ubiquitous expertise', the expertise needed to exist as a citizen in a particular society; 'technical expertise', which has five levels, including 'interactional expertise' and 'contributory expertise'; 'meta-expertise', which is expertise about the nature of others' expertise; and so on. Meta-expertise is what people use to judge between competing experts and competing expertises. Meta-expertise is what the citizen uses to decide whether tea-leaf reading is as good as astronomy when it comes to predicting the appearance of comets and so on and so on. As most of us see it, meta-expertise is flawed when the views of distressed parents, or celebrities, are preferred to views of medical experts and epidemiologists if vaccination decisions are being made. If computers are to act like humans they are entitled to get these things as wrong as citizens sometimes do, but they will have to do much better than random; somehow they will have to fit

into society in this other way – they will need to make similar
meta-expertise judgements as the majority of sensible citizens
in a society most of the time if not all of it.[100] Intelligent
computers will have to reflect the way social groups work.

Every society is made up of groups within groups within
groups, overlapping and cross-cutting in a multidimensional
way. Since each group and sub-group, at whatever scale, has
the same characteristics – one can only become a member
through socialization, absorbing the tacit knowledge and
learning the specialist language – this is called the 'fractal
model' of society. Figure 8.1 is a sketch of the UK in these
terms but, of course, in only two dimensions.

In Figure 8.1 the ubiquitous expertises are at the top. The
entry for pavements stands for citizens' general knowledge
of how to navigate society, represented by knowledge of how
close to walk to others on the pavement (sidewalk), depending
on how crowded it is, something that will vary from society to
society; bananas refers to UK citizens' objection to European
legislation on the shape of bananas (possibly mythical, but
UK citizens were pretty sure they knew who the experts on

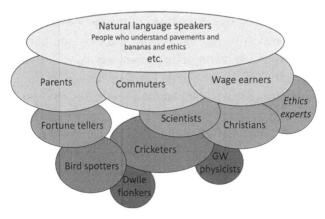

Figure 8.1 *The fractal model of society*

bananas were); the entry for ethics is interesting because without ubiquitous expertise in ethics we would not feel justified in having a criminal justice system – people have to know what is right and wrong if societally sanctioned punishment for doing wrong is to be legitimate (Ava, Samantha and HAL lacked this ubiquitous expertise and weren't good at knowing right from wrong). The bottom entry, 'Dwile flonkers', refers to a group of players of a made-up game which probably does not quite comprise a social group and so maybe slips off the bottom of the fractal; dwile flonkers are there just to show where the fractal comes to an end. The other entries, however, all represent sub-groups within a society like that of the UK, and the fractal idea is also useful because each of these sub-groups has a similar structure, whether big or small. To each group we could apply the Periodic Table of Expertises and describe what counts as the ubiquitous expertise for that level (it would count as a specialist expertise when seen from above), then describe what is the level-specific meta-expertise, and so on. This is the kind of structuring of expertises that computers will have to understand if they are to reproduce human abilities. In this chapter I am going to look at an aspect of the way meta-expertise works in the case of gravitational-wave physics. The same exercise could have been carried out in respect of some other level of expertise, but in this case I happen to have rich details.

Thus, I am going to use my own project on gravitational-wave physics as the prism through which to observe the process of understanding a strange culture and keeping abreast of it as it changes. My project has been going since 1972 with a particularly intense period from the mid-1990s to the mid-2000s, when I spent more time in the company of gravitational-wave physicists than any other specialist group. I am going to concentrate mainly on a particular couple of days of interaction with gravitational-wave physicists that took place at a conference in Budapest at the beginning of September 2015 to draw out some of the ways that humans interact in a field like this, using myself as a kind of experimental probe. This meeting took place a couple of weeks before the momentous first discovery of a gravitational wave – that observation being on 14 September 2015, with the announcement to the world being at a series of press conferences on 16 February 2016.[101] What is going on in this chapter involves my drawing on certain special experiences that contribute to my right-hand-side-of-Table-2.2 expertise (p. 30). Some of the observations I report arose because I had this book in mind during the Budapest meeting and arranged certain conversations that might not otherwise have taken place. It was lucky because a couple of weeks later everyone's attention was captured by the momentous discovery.

The fact that I can use these few days in this way rests on my embedding in the society of gravitational-wave physicists, which had been accomplished over the multi-decade earlier life of the project. In particular, the physicists had come to trust me and they were ever ready to talk to me when we sat together at lunch or coffee or some more deliberately arranged meeting. The trust had been built up by my never betraying any confidences over the many years of contact, by my extended and assiduous contact with the field, and, perhaps, by my passing a well-publicized Imitation Game as a gravitational-wave physicist, demonstrating that I had sufficient technical understanding of the field to be worth talking to.[102] The documentation of trust in the last sentence is not just a report of how I came to be embedded in the field, but is an observation of how humans work together; it needs to be taken into account when we are thinking about whether machines can mimic humans. Human interaction in a knowledge-creating field like this is based on trust! What is the computer equivalent of trust? Maybe the emotional

failings of Ava, Samantha and HAL (Chapter 1) do have
something to teach us after all; how does one come to *trust* a
computer? As it is, we can't rely on them to keep our secrets
but have had to build a great industry of anti-hacking software
to safeguard them. That, of course, is now, not the future.

Lessons from gravitational-wave physics

The lessons about human interaction that were learned, or
perhaps merely illustrated, by my multi-year induction into
the field and my two days of directed observations are sum-
marized in Table 8.1.

The sociologist is reassured by symbols of trust expressed
by respondents: arriving at the Budapest meeting I note that
I am greeted by smiles and waves and inquiries about my
health, and whether I will be at the forthcoming series of
meetings and so forth – I'm now a well-known acquaintance
in this community, pretty well a member of it. Once more, at
lunch and dinner I sit with the community – not infrequently
with those at the very top of the organization, since I am now
one of the longest-standing members.

The obligation mentioned in line 1a of Table 8.1 is ever
present. Right at the beginning of my fieldwork on this topic
in 1972, I wrote to the US scientists I wanted to interview,
telling them that I was coming to America for my PhD work
and that I would phone them when I arrived to arrange an
interview. I hoped they would feel obligated to help a PhD
student who had already spent resources on transatlantic
travel. It worked pretty well and no one refused, though one
took a bit of persuasion. This, of course, was just the entrée.
But at difficult moments later in my fieldwork, I knew that
I was being tolerated because I was a fellow academic and
academics feel an obligation to each other.

To see an illustration of the relationship between trust and

Table 8.1 How humans learn from each other and develop new knowledge

#	DESCRIPTION	GLOSS
1	Trust and	*Trust consolidates and bounds human groups*
1a	Obligation	*Obligation to another party of similar status encourages trust*
1b	Originality	*Original thought showing technical fluency encourages trust*
1c	Expertise	*Trust comes with technical virtuosity and clear commitment*
2	Fluency and linguistic practice	*Cultural/technical understanding is acquired through active engagement with the conversation just like practising any other skill; like any other skill it has tacit components*
3	Social contiguity and	*Face-to-face meetings are considered vital to humans*
3b	Simultaneity	*They are efficient because many interchanges with a number of people can be conducted almost simultaneously with each person being aware of all the interchanges simultaneously*
3c	Body language	*Implicit meanings come with smiles and body language*
3d	Privacy	*Interchanges conducted among small groups with secure boundaries develop new knowledge*
3e	Serendipity	*Social contiguity allows for serendipitous interactions*
3f	Familiarity	*Meeting people gives a sense of who you are dealing with which cannot be obtained through text*
4	Forceful disagreement with trust	*Ideas – new patterns – emerge from disagreement, often forceful disagreement. That is why the sociologist trying to gain new knowledge must not be deferent. Face-to-face conversation is vital if productive forceful disagreement is to be managed by body language and trust.*
5	Commensality	*Commensality (eating and drinking together) helps combine disagreement and trust*
6	Not meaning what you say	*To grasp that a conversational turn is meant to be provocative, not literal, needs an understanding of context; it is a feature of polimorphic actions*

originality – line 1b – we go back to the gravitational-wave Imitation Game, and how it was that the author was so successful that of nine GW physicists faced with the dialogues, seven said they could not tell who was who, and two chose me as the real physicist.[103] This surprising result came about in good part because I did not know the published answer to one question, so had to think it up 'on the fly', giving an answer that was right and not to be found in any known publication. It was this originality that convinced the judges that more expertise was being displayed in my answers than in those of the genuine physicist, who did reply with the published answer.

Line 1c is logically prior to 1b in that the originality discussed under 1b is what leads to the attribution of expertise, and it is expertise that is another route to trust. Genuine experts are trusted because they are seen to have a deep commitment to what they are investigating; shallower commitment would be revealed by lack of expertise. Those dipping in and out of a field from the outside, gaining only superficial understanding for the sake of a quick return, are less likely to be trusted, because the relative weighting of reward to commitment is not conducive of trust. Almost too obvious to be worth mentioning are the other indicators of commitment: long-term involvement with a community, especially if it is demanding of time and effort (for example involving a great deal of travel), and, still more obviously, the preservation of confidences during that long-term commitment.

Let us remind ourselves why we are engaged in this exercise. First, we want to explain what will be needed by Level IV intelligent computers – computers that exactly reproduce the way humans work. Level III computers – computers that do the same job even if the mechanism is different – may well be no different in these respects. If any computer is to reproduce the way humans work it is likely that they will need to engender trust in all the ways just described in order to be allowed entry to specialist groups. It seems arguable that computers will need at least enough in the way of bodies to be trusted in the way humans are trusted. This does not mean they have to be athletes, given that many seriously disabled humans are trusted, but it does require demonstration of the emotional timbre that engenders trust, and this often comes with bodily activity and body language. Though, contra Dreyfus, it is not a necessity, given interactional expertise, that an individual have a human-like body to grasp the conceptual world of the fully-abled, it does seem that individuals need to be engaged in some kind of bodily interaction to create the conditions that will allow them to gain interactional expertise through linguistic interaction with other humans.[104]

There is, however, a contradiction here with what we found in Chapter 1. Even though it turned out that the trust was not justified, Ava, Samantha and HAL were all trusted by their human companions even to the point of falling in love in two of the cases, and we argued there that this was because they were fluent users of the language. Of course, we can all fall in love with perfectly fluent people who shouldn't be trusted and it happens all the time, so why shouldn't we imagine that this small group of gravitational-wave scientists would come to trust a computer in their midst, more especially one that could demonstrate a considerable degree of technical proficiency? The answer is that we *can* imagine this happening and that is why we need to remain aware of the limits of computers until they reach the point when trust can be justified, rather than granted mistakenly to the wrong kind of entity: we don't want to do the equivalent of falling in love with Ava or Samantha, or handing over technical command to HAL. It may be that the majority of computers, if not all, will have to interact bodily to engender deserved trust – who knows.

Note that everything that has been said so far is about

gaining trust in order to gain entry to a specialist group in order to assimilate and contribute to their knowledge. *This is opposite to the way computers currently gain knowledge via deep learning.* The apogee of deep learning is having computers plugged into *all* of the world's knowledge, whereas the world described here is one of specialists fiercely guarding their boundaries, not for reasons of propriety, but because they know that only members of the specialist group are capable of interpreting the knowledge in an appropriate way. The art of doing science includes careful policing of the boundaries of specialist groups.[105]

Much of what follows expands on these points with more description of the author's interaction with gravitational-wave physicists, particularly at the Budapest meeting. This reveals the full richness of what computers would need if they were to reach Level III or IV, including how genuine trust is earned and how social context works in human interaction.

Fluency and linguistic practice (line 2)

As the gloss corresponding to line 2 in Table 8.1 explains, cultural/technical understanding acquired through speech is acquired through active engagement with the conversation, just like practising any other skill; as with any other skill it has tacit components. To learn from conversation is not a matter of being instructed – that is too unidirectional. Instead one must be engaged and that often requires confrontation, such as arguing with the conversational partner by, perhaps, confronting them with the arguments of other scientists. Arguments can be found at the frontiers of a science that is still forming through disagreement. An article of mine, published in 2004(b), explains the progression in this kind of growing comprehension, or acquisition of interactional expertise:

As a sociologist of science you essay research on a new specialism and you initially understand neither the banter nor
the technical terms. After a painful period, if you are lucky
you begin to pick up on the inferences in others' conversations and eventually you begin to be able to join in. One day
a respondent might say in response to one of your technical
queries 'I had not thought about that', and pause before
giving you an answer. When this level is reached respondents will start to be happy to talk to you about physics and
even respond generously and with consideration to your
critical comments. Eventually people will become interested
in what you know, not as a scientist in your own right, but
as a person who is able to convey the scientific thoughts and
activities of others. If you've just come from visiting scientist
X you may be able to tell scientist Y something of the science that X is doing . . . What were once 'interviews' then
become 'conversations' that can be interesting and occasionally even useful to both parties. What also happens in a
conversation is that by occasionally anticipating a point your
partner is about to make you can speed things along. You
might also verbally fill in some gaps that might otherwise
be forgotten. You can recognize jokes, irony and when you
are having your leg pulled (though, in the nature of things,
interactional competence does not allow you to recognize
lies). When you get good at it you can even take the devil's
advocate position in respect of some scientific controversy
and maintain it well enough to make your conversational
partner think hard. (Collins 2004b: 104)

Notice the importance of understanding jokes, leg-pulls and
banter.

This approach reflects the way scientists gain their own
understanding. Of course, they go through a period of formal
education where they sit in classes and take notes, and they

take instruction from their superiors, but if they are ever to reach the research front they will need to take an interactive part in the discussion: they will have to ask challenging questions and put forward challenging ideas. Somewhere along the line, deference has to give way: being rude to your betters and seniors is never going to work but if you are not ready to challenge them, and banter with them, the knowledge will never become your own. Gaining real knowledge is not learning from text or even spoken instruction; it is a matter of forceful engagement with the knowledge, either through practice or argument. One cannot become a carpenter by reading everything there is to know about carpentry or even listening to everything there is to be said about carpentry – one must do some carpentry. In the same way one cannot learn *everything there is to be said about carpentry* (gain interactional expertise in carpentry) just by listening to everything that is said; one must do some saying – engage in the *practice of* carpentry conversations.

Social contiguity (line 3)

Another feature of human interaction, as opposed to computer interaction, is social contiguity. People work better when they are close to each other; there is no equivalent in the computer world. Once, I was at one of these meetings in Australia and noted that Gary Sanders, the LIGO Project Director at the time, had flown to the meeting from California – a round-the-clock trip – and left after one day to return, by another round-the-clock flight, without spending a night in the southern hemisphere. Sanders felt his personal presence at the meeting was important enough to justify this kind of exertion.

In this community the use of meetings has evolved over the years. When gravitational-wave physics turned into a big science with a large community that had to be held together, there were more meetings than there are now, as people got to know each other and mutual understandings were established. I went to a large proportion of these meetings, maybe half a dozen a year. Now they have been cut down to two main meetings per year. The physicists are pretty sure that fewer meetings would not work. They have also tried more meetings, but found they were too time-consuming and had diminishing returns. But they claim that two is enough only because of the existing face-to-face tradition that was developed at a time when there were many more meetings.

In Budapest I asked scientists why they came to meetings and did not just stay at home and communicate over email and the like. This group, bear in mind, are among the most computer literate and best internet-connected people on the planet. A series of reasons emerged (pseudonyms are used):

> *Beta:* It's an indication of how much better face-to-face meeting is that we will make all this effort to be here. For example, I am smiling and over email someone might get mortally offended by the same remark if they could not see the smile. And the iterations happen over a very short time span so it is very efficient.
>
> *Gamma:* I'm here even though it is taking a week when I am not being as productive as I would be doing normal research, so it must be worthwhile.
>
> *Alpha:* There are certain limitations to email communication: it's to overcome those limitations. There is efficiency to face-to-face communication that you just can't get any other way.
>
> *Gamma:* Well, we've known each other for seventeen, eighteen years or so, so it [trust] is not much of an issue but newcomers might well find it important.
>
> *Alpha:* I guess we have a lot to bank on.

One day I chance on Beta in the corridor (emphasizing the serendipitous aspect of physical contiguity – line 3e), and he mentions that the format of the meetings has changed and coffee and lunch time are now much longer at the expense of formal meeting time: 'this is so we can chat to each other'.

The scientists stressed the speed and efficiency of face-to-face interaction: questions and responses were exchanged at the speed of thought, so that a long sequence could be completed in a very short time and with a number of people. Compared to this, emails and other more distanced means of interaction were clumsy and slow. Furthermore, you could learn much more about what people really intended from talking face to face:

> *Delta:* Here you can just go round a bunch of people and in ten minutes the whole thing [a new collaboration] is arranged and you can tell immediately whether the people are enthusiastic, how interested they are, how much commitment you are going to get from them just from their reaction, which you would have no idea about over email. In email [you get]: 'that's great! I'd love to do that' [but it does not mean anything.]

So we have the importance of smiles (line 3c), proximity of the members of small groups (line 3b), and the obligations implicit in face-to-face agreements that are not found in electronic media.

For novices there is an additional value (line 3f):

> *Beta:* It is very useful for students to come to these places and learn what the people with certain ideas look like; I think this helps enormously. It helps to associate an idea with the person involved. I have seen beginning students trying to hold everything together and then they come to one of these

meetings and suddenly it all falls into place for them because they can see the people who have the ideas. It is also much better than trying to introduce yourself to someone over email – you write the email and say, 'I am so and so', and they may not appreciate the relevance of the student's work to the work of the group from just an email. Having others around to back the student up or to help put their work in context can be important, and this can be done better face to face rather than through less nuanced email.

Incidentally, in none of the many email and other kinds of online conversations that I engaged in over the years (including telecons) was video ever used and no one ever suggested that video would be helpful. Skype is nice for communicating with your grandchildren but does not seem even to approach the effectiveness of proper face-to-face interaction. None of this proves that video might not enhance distance communication a little but the distance between video and face-to-face is so huge that the increment it might add is not worth mentioning. What is a useful increment to telephone conferencing, which seems to be widely appreciated, is the means to maintain a simultaneous text interchange.

Forceful disagreement with trust for both learning and creation (line 4)

Now let us return to the question of needing to be confrontational to learn: these responses were unprompted apart from my initial remarks ('Why do you come to these conferences? Why don't you just stay home and communicate by email?'). Consider how this way of learning interactional expertise compares with the way deep-learning computers do it.

Delta: Science is a critical business but also learning is. You

don't learn things just by reading, you have to engage
with it. If you are able to say, 'I don't agree with that', or
'I don't understand that', or 'I think that's wrong', then
you understand it better and maybe new ideas come up.
But if there's someone there, some professor, who knows
it better and says 'it's like this – OK?' [Then you feel] I
don't understand it – I'd better go and read a bit more
about it: that's not useful.

There is a skill to disagreeing without offending people:
a skill to staying just on the right side of rude, and it is a
skill that will depend on all manner of subtle features such
as how long you have known the other party, what is your
relative status and, perhaps, whether the situation is formal
or informal – maybe with a few drinks taken over dinner.
Commensality (line 5) is a regular element in the build-up of
trust – one cannot have anything much more dependent on
the body than eating and drinking together. And using these
things properly depends on body language.[106]
I sat in on a heated argument between Alpha and Beta over
the usefulness of a new form of statistical analysis, which Beta
claimed could take account of a certain kind of 'noise' and
Alpha insisted could do nothing for the work of the group.
Voices were raised and the parties often spoke across each
other with the vigour of their arguments, but there was never
the slightest chance that this would lead to falling out; both
parties realized they were trying to reach a point of mutual
understanding and agreement, and this was best achieved by
disagreeing in the most forthright way – forcing the other
party to clarify so as to be as persuasive as possible, and thus
bringing out the key points of all parties' arguments both for
themselves and the opponents.
Another example took place in the same interchange. Beta,
at one point during lunch, shouts out forcefully that 'you

should never do time-slides' – time-slides being the fundamental method in gravitational-wave physics for establishing
the background noise against which the significance of signals
is estimated. But it turns out Beta does not mean it! And
Alpha and Gamma know he does not mean it (line 6). He
means that what you really want is to understand the noise in
a theoretical way – that is the proper way to understand the
significance of the signal. But all three of the speakers realize
that at the current state of the art you cannot understand the
noise in a theoretical way, so if you want to understand it at
all you have to do time-slides. What Beta is saying is that, to a
physicist, time-slides are an unsatisfactory pragmatic solution
because there is no other way to get at the noise. Alpha, Beta
and Gamma know that Beta is affirming that time-slides are
not what they should be doing as physicists but this is the best
they have. Beta expresses this by saying, 'you should never do
time slides,' and the physicists understand this, but I don't – I
am just hearing those words and interpreting them as one
would most of the time. The physicists have to explain it to
me. This, of course, is a perfect illustration of polimorphic
action instanced in the way we understand language. It is just
like when I greeted my partner as 'you bastard'. We will see
another instance shortly.

On the same topic:

Delta: When the collaborators are in different places you
can have tensions and misunderstandings can build up. And
often you get really frustrated with someone and I think they
are doing things wrong and they misunderstand things and
maybe we're just not going to continue working together
because this is not going well. And then you see them in
person and you have a beer and you chat about things
and then it's all fine. If you don't have that every so often,
things can get very complicated – for no good reason – just

misunderstanding – paranoia . . . Maybe they're trying to get to a result before you, or they've got some hidden agenda. To some extent we're all trying to do this, we're all trying to get ahead, but the fact that you are working together and you respect each other and you want to continue working together – that doesn't come across in a telecon[ference] or certainly not in emails . . . I think people who don't go to meetings and stay away – I think they get wound up in their own world.

Returning to the previous point:

Epsilon: I think you also have to understand the role of the people you're talking to. If they are above you, you tend to say only what they expect and what's absolutely expected by everyone, and if they are at a similar level you can make more provocative statements. But it takes a while . . .

Epsilon: Often this is a great way to trigger really new thought and not just walk along saying the same things that everyone is saying, just in a different way. Sometimes you want to say things that are pushing it a bit harder just to see how other people react to it. And that's how a new thought emerges. You don't want to come across as being a provocative type and too stupid to understand the common consensus, but you just generally want to try out things.

Alpha: I can certainly recall occasions when random conversations over lunch and coffee have triggered some insights or ideas. Chance conversations can generate these things.

Epsilon: I think sometimes it's useful to have this role-play where you say something and I deliberately try to find arguments against it, though my instinctive reaction

might be to agree with you but doesn't help, so I try to be
critical and I say 'no – why isn't this and this?'

Delta: Some of the very best discussions I've had – some
of the best interactions and experiences in science are
in a small group of people when you can say 'no that's
wrong' and you can argue about it, and there is a sense of
trust the person doesn't think you're an idiot: they know
you're smart and you're confident and you're asking a
real question, and they're honestly trying to explain it to
you and you're honestly trying to understand it, and from
that ideas flow and one of the most interesting results,
the XXXX thing, came from a week of – there were a few
of us – and at the beginning of the week someone says, 'I
think the thing looks like this,' and then someone says, 'I
think it looks like this, no it's not like that,' and back and
forth and arguing, and within a week we'd found something new that we truly didn't expect at the beginning.

How do we keep these confrontations friendly? The point has
been made – it is by commensality along with body language.
Eat, drink and be sociable; disagree while using subtle cues to
demonstrate the absence of enmity.

Body language lost

All the points in Table 8.1 have now been illustrated. I have
mentioned that I was a bit slow picking up on the meaning of
Beta's forthright statement: 'you should never do time-slides'.
One more visit to the problem will also illustrate how closely
linked are understanding body language and understanding
the technical culture. My contact with the community has
been more irregular than it was up to the mid-2000s and here
I show how I suffered from it.

Over dinner on the Tuesday (commensality again!), I find

myself sitting next to two of the most senior scientists, one of whom I'll call 'Doug'. Our conversation turns to 'Big Dog' – a 'blind injection' event. A blind injection is an artificial signal surreptitiously injected into the detectors to cause the scientists to rehearse their analytic procedures.

A main theme of my 2013a book concerning the blind injections is my disappointment that the statistical significance of Big Dog never quite reached the conventionally agreed level required for a 'discovery'. The gravitational-wave community have so far adopted the current convention of high-energy physics, which means that to announce a 'discovery' one must have a statistical significance of at least five standard deviations, while between 3- and 5-sigma the right claim to make is only 'evidence for' a discovery. I argue in the book that the community should have 'removed the "little dogs"' so as to increase the statistical significance of Big Dog. This is an esoteric and philosophically puzzling aspect of gravitational-wave signal analysis. In the book I try to provide a little proof that the little dogs should have been removed.[107]

Now Doug says to me something along the lines: 'of course the Big Dog signal was statistically significant and it is obvious that you must remove the little dogs, otherwise you get a step-function in the noise analysis and that can't be right'. Doug says this with real force, as though only a fool would not understand it, and I gather from the way he says it that the community is now uniformly of this opinion. I am somewhat 'thrown' by this but, at the same time, quite pleased because it suggests that I was right about one of the only bits of physics I have ever tried to 'prove'. It seems to me that things must have moved on while I have been distant from the field and it is now agreed that little dogs should be excised!

But then I talk to 'Quince', another very senior physicist, with whom I had an extended argument about the little dog point a few years back – he, at the time, being staunchly of the view that little dogs must not be removed. I ask Quince what he makes of the fact that it is now generally agreed that the little dogs should be removed. Quince looks at me quizzically and assures me that he is still very much of the same opinion as before. Quince tells me that he is the Chair of the Detection Committee which is charged with making this decision and it is far from settled.

I ask around and find that Quince's view reflects the current state of affairs and what I heard Doug saying was wrong. Had my understanding not degraded with my slightly more distanced relationship with the community, I would have known that the discussion was still ongoing rather than settled, and that the opinion of the wider community had not changed after all. But that is not the deep problem; much more revealing for the purposes of *this* book is that I had misunderstood Doug's body language. Doug was telling me with the utmost force that it was obviously silly not to remove the little dogs, but in doing this he was not intending to tell me about the consensus in the community; he was telling me his point of view and telling me that he really thought he was right. Doug was not revealing the truth of the matter but trying to create a bit of the productive dissent that has been discussed above: 'It's obvious to me that I'm right and I challenge you to disagree with me!' I missed it because I am no longer quite keyed in to all these nuances – I am not fully apprised of the context of what is being said – and interpreted it as a declaration of a certain factual and social state of affairs instead of an invitation to argue. One sees then that even straightforwardly and forcefully uttered statements of 'the truth' are unreadable without the social context: that is how human communication works!

I sent my interpretation of events back to Doug, and on 12 September he told me, 'You have it about right.'

Interim conclusion: What this study of human interaction means for AI

Here, then, we have some illustration of humans learning about and adding to the understanding of a new area of the physical world, and how this knowledge is moderated by all kinds of face-to-face activity, including commensality and body language. What has been described are the ways some of the links between the top layer of neurons in Figure 7.2 (p. 134) and the rest of our brains work. As stressed throughout this book, this is the big problem of AI: the articulation of machines with human culture – the methods of absorbing culture and contributing to culture, all of which turn on understanding social context. Here we see the way this articulation happens within small groups.

But maybe there are artificial ways to engage with culture that do not involve what is described above. After all, some of what humans do, as described above, may be a necessity borne of their deficiencies: their slow brains and limited capacity for extended work and concentration. Maybe computers would not need to limit the amount of information they process because they are capable of processing all of it in a way that humans cannot. There could be some truth in this claim, but mere quantity does not seem to be the problem. What humans are doing with the strong boundaries they erect around who is admitted to circles of expertise, and with the methods they employ to acquire and extend bodies of expertise, has more to do with quality rather than quantity. There is *too much* quantity and too much in the way of contrary opinion in the world. The social interactions described here make up the method humans use for limiting and assessing this overwhelming burden of information and contrary ideas, and reaching conclusions in spite of it. It is hard to see how computers can mimic these methods.

The stubborn but strangely overlooked problem of AI

The position argued throughout the book is that the problem of AI is engagement with society: computers need to find the means to understand the world of the embodied collective and the different worlds of practice within sub-communities. The new, pattern-searching AI and its engagement with the internet has resolved it to some extent. What is being argued here, however, is that this is only a partial resolution; it simply brings the frontier of computer-understanding nearer in time to current human understanding, but it doesn't allow it to merge with the developing flux of human understanding and consensus formation.

Maybe the argument has become confused or confounded by the ideology that, if not self-consciously endorsed by most AI researchers, does obscure what is going on when AI is presented in the public domain. The ideology is that we are all individualistic machines shaped by evolution and capable of learning all we need in the course of our own, free and pioneering, confrontations with the world; the political resonances are strong. In contrast, the way humans and computers interact with society has to be at the heart of the quest to develop general artificial intelligence. Humans get nearly all their knowledge from the top, and like computers they get a lot of this top-down knowledge in some kind of implicit way – from the distribution of words, the intonations and silences, and the body language in which they are immersed from birth.

Kurzweil thinks that computers will become more intelligent than us because:

(i)   machines can share their knowledge at extremely high speed, compared to the very slow speed of human knowledge-sharing through language

(ii)   Nonbiological intelligence will be able to download
       skills and knowledge from other machines, eventually
       also from humans
(iii) Machines will have access via the Internet to all the
       knowledge of our human-machine civilisation and will
       be able to master all this knowledge. (2012: 26)

With its mastery of huge numbers of internet pages, the
Jeopardy! winning computer, Watson, shows what can be
achieved with quantity of information and speed. But this
is not how human knowledge works. Some things can be
achieved with quantity but mostly the problem is interpretation – not quantity, but quality. Quality is tied up with trust
and, in humans, this depends on features of face-to-face communication as described in Table 8.1.

An incident from the history of gravitational-wave detection, which will be referred to again in later chapters, further
illustrates the importance of credibility when assessing facts.
Joseph (Joe) Weber is the pioneer of the terrestrial detection
of gravitational waves. In the late 1960s and early 1970s,
Weber claimed to be detecting gravitational waves, but by
1975 very few people believed his results (as explained, it is
now almost uniformly believed that the first detection took
place in September 2015). In 1996, Weber published a paper
in a physics journal, claiming to have a found a correlation
between gamma-ray bursts and the gravitational waves he
believed he had discovered in earlier years. On the face of it,
this was a potentially Nobel Prize-winning finding. But when
I investigated the opinions of a large number of gravitational-
wave physicists, it appeared I was the only person in the
community to have read the paper. The 'technical' judgement
being made was that Weber's credibility had now fallen so low
that this was a 'non-paper', in spite of its perfectly respectable
appearance. As far as the internet and any reader outside

the specialist community was concerned, however, it was a new piece of scientific information. With the help of Paul Ginsparg, the founder of the physics preprint server, arXiv, we showed that Weber's paper would pass arXiv's automatic software tests for 'fringe' papers. But Weber's 1996 findings were not data: data only become data when they have been filtered through the sieve of human credibility.[108]

Small groups, trust and the body

Returning to aspects of secrecy and the way small groups control their boundaries, one characteristic of human inter-action, at least at the frontiers of science, is that there are well-defined boundaries to the groups taking part in a debate. The problem to be solved is not one of gathering all available information, but that of restricting information to what is believable and limiting knowledge contributions to those who are believable. There is no formula for this; it is something that emerges through the interactions of expert groups.

> *Delta:* Part of being an expert is understanding whether what's being said is legitimate, what result people believe and what they don't believe and what's out of fashion, and you can't get that unless you are around other people and you're seeing their reactions. You're sitting at a table at lunch and you bring out some paper or some theory and you can see how everyone reacts. And if you're writing letters to them asking them [that won't happen] . . .

It may be that we learn to trust people close to us who give us food or eat with us from the very early days of our life. There is nothing in our early socialization that encourages us to trust large groups – on the contrary, we are warned against trusting strangers; perhaps we carry all this into adult life and this is why science works as it does. (Size of groups is important in other kinds of human interaction as the shock of Samantha's 641 lovers indicates. Note also how suspicious we are of societies ruled by 'Big Brother' dictators, where loyalty to the leader is confused with loyalty to a family member.)

If trust in science is built up in small groups through face-to-face interaction; if the serendipitous effects of proximity of one body to another are a crucial input to the building of knowledge and the recognition of new patterns; if body language – smiles and the like – are a vital component of understanding what is said and how it is said; and if eating and drinking are vital components of the way the top layer of neurons interact with one another, does this mean that the body is a necessary condition of the understanding after all? Insofar as the answer is 'yes', the body discussed here is not a condition for forming the individual brain's model of the physical world; it is a condition of the individual's interacting with other brains to form a mutually agreed picture of the world. This body has to do with interactions with other bodies and minds, not with the physical world.[109] Second, as has been pointed out, the body cannot always be necessary even for this modified role because there are people who cannot manage this kind of interaction yet still fulfil their lives within scientific society.

The variations in the need for social interaction are made possible by the division of labour in science: not every scientist has to be an expert in every aspect of a field. After all, 'unintelligent' computers now do a lot of the mathematics and complex calculation that were once thought the prerogative of brilliant scientists, so, those brilliant scientists were doing what could have been achieved by *unintelligent* computers. But this does not mean the entirety of science could be done by unintelligent computers; that would go back to the simple BACON model. For human science as we know it to work, a good proportion of the community must have bodies that can engage in commensality and all the rest of it; just as in the previous discussion of the body, at the collective level the body is essential. So if our ambition is to mimic all the kinds of scientists there are – to reproduce 99.9 per cent of human abilities found at the frontiers of science – there must be a way for machines to eat together or some equivalent; some way of creating the trust that humans depend on in their small groups while closing the boundaries to all but the trusted insiders. And even if you are determined to preserve the simple model of science that would allow a program like BACON to work – however much you are determined to ignore the time it takes for the oil in the lava-lamp to reach the surface – the fact that there are many kinds of human knowledge generating activities outside of science means that the problem will not go away.

In the case of the previous discussion of embodiment, it was argued that an individual computer could learn to understand anything *already known* that was associated with collective practice, solely through immersion in the appropriate linguistic discourse. Here, however, it is being argued that some of the kinds of discourse that are needed to *create new* knowledge cannot be engaged in unless the majority have the right kind of body and, given the propensity and necessity for privacy and control over the boundaries of scientists' discursive space, some kinds of knowledge cannot even be *learned* without the right kind of bodily engagement by nearly everyone who wants to learn (if not absolutely every individual).

9

Two Models of Artificial Intelligence and the Way Forward

We are concerned with whether AI will amount to general intelligence in the foreseeable future because, if it won't, this needs to be widely understood so that we don't enslave ourselves to stupid computers. We have found that to understand the highest ambitions of the AI project we need to think in terms of six levels. The levels are distinguished by the amount of asymmetric repair that we humans do: by whether we are trying to mimic or exactly reproduce human capabilities; whether we are looking at isolated machines that are parasitical on human cultures, or at autonomous groups of intelligent machines; and by whether the autonomous machines have bodies, and whether those bodies are human-like or alien. AI is currently at Level II, *asymmetric prostheses*, while nibbling at the edges of Level III, *symmetric prostheses*. The gap between these two levels is crucial, with Level III AIs being able to pass the most demanding of Turing Tests and therefore satisfying the criterion of general intelligence that is the main focus of this book. But it may be that Level III cannot be attained unless an awful lot of the workings of the human mind and human society are incorporated into computers. In other words, the attainment of Level III may turn out to depend on going much closer to Level IV than is discussed in the preceding chapters; this is still to be resolved.

Another way to summarize the argument is to frame it in terms of two models of artificial intelligence, a framework that relates to the two models of science we have discussed: the 'monotonic model' and the 'modulated model'. With the monotonic model, there would be no need for machines to draw on human culture in order to achieve the highest reaches of human attainment – which is taken to be scientific understanding. The modulated model, on the other hand, requires machines with general intelligence to be embedded in society because even science is embedded in society. The main question of this book concerns whether machines are becoming embedded in society and, especially, whether this is true of deep-learning machines drawing on indefinitely large databases.

The monotonic model

Under the 'monotonic model' all cleverness is the same – more silicon nodes working faster mean bigger, quicker, artificial brains, and that means cleverer artificial brains, all adding up to a universal, internally consistent, hyper-Wikipedia. As the exponential accelerates, artificial brains will reach the level of cleverness of human brains and then surpass them – surpass them to an unimaginable extent.

This monotonic idea of cleverness is supported by a view of science as the accumulation of true, clear, universal and undisputed knowledge. Under this model our idea of a brain and our idea of a scientist are one and the same. For the knowledge to be universal it must arise from the world, not from disputing humans – it has to resemble logic or mathematics (simply understood) and therefore it can only have been generated from the bottom up because it is only the bottom – the physical world we encounter – that is sure to be the same everywhere; science works on an elaborated geon-like model. This is the simple model of science that supports BACON and its supposed deduction of Kepler's Laws. It is thought that, insofar as social scientists have disputed the monotonic model, all they can have done is describe a lava-lamp – truth reaches the surface a little slower than we thought, but that is because humans are frail. Faster computer-brains will heat the liquid so that the oil will reach the surface more or less instantly, without the need to worry about all the social interaction that takes place in science and that the social scientists have studied; if everything speeds up and the oil reaches the surface instantly, there will be no social interaction worth mentioning. Under this model, *true* human knowledge arises from the bottom up with clumsy humans being, at best, the facilitators. It is due to evolution that even clumsy humans reach the right conclusions because only those reaching the right conclusions survive. More recently, free-market capitalism has been stirred into the mix.

The modulated model

The alternative 'modulated model' is described in this book. One version starts with the sketch of the individual brain as described by Kurzweil – hundreds of millions of hierarchically arranged pattern recognizers. But the brain, human and artificial, is thought of as starting out with potential only, and therefore nothing about the way it will exhibit cleverness is assumed; Ava, Samantha and HAL, the artifictional intelligences, have been created by film-makers to be as they are and there is nothing in the substance of their brains to make them that way. The dominating question for the modulated model is how the brain, human or artificial, gets filled up, once we no longer assume that left to confront the world by itself it would automatically fill up with true scientific knowledge. Sociologists are led by their *ideology* to think mainly about top-down filling processes, but here the ideology has been modified to admit that the world would make no sense without some bottom-up filling – the geon-like model or similar. This will lead to some, minimal, universality in conceptual and perceptual structuring of the world. But this is just a necessary condition for there to be any knowledge at all, while the interesting processes that give rise to all the differences in the way the world is categorized are top-down. When it comes to the analysis of existing AI projects we find that the relationship between bottom-up and top-down knowledge-filling is not clear, either in fact or in the way the work is described.

Important to the sociologist is the work on developing an improved understanding of science that has been done, starting in the 1970s. But even if this is not believed, or thought only to reveal what happens as humanly slow lava-lamp science stabilizes, it makes no difference. This is because the lava-lamp does not stabilize at all in the case of the large majority of sciences that, unlike the movements of heavenly bodies and sub-atomic particles, involve complex processes. Insofar as a lava-lamp represents what happens in non-scientific cultural enterprises, then the stabilizing mechanism is certainly not driven from the bottom. That is why human cultures and sub-cultures – even sub-cultures within science – have widely differing views of the world: the brains are filled with many different knowledge substances. Most of us believe gravitational waves were first detected in September 2015, but some physicists working in universities do not believe it (see Collins's *Gravity's Kiss*), and this is before we look at climate change, econometric modelling and the like.

The modulated model of the brain can be built on top of the monotonic model by adding another level of quasi-neurons above the hierarchy of pattern-recognizers that, in the Kurzweil brain, make up the neocortex. The extra layer of quasi-neurons comprises all the other brains with which an individual brain interacts. This is the 'societal layer' of neurons and it reinforces or vetoes patterns that are proposed in the rest of the neocortex in ways that are similar to the way in which the higher level of the individual's neorcortex reinforces and vetoes – though the mechanism is sparser, slower and more complicated.

The body revisited

What about the brain and its relationship to the body? An influential critique of 'good old-fashioned AI', or GOFAI, held that computers must have bodies to be intelligent. Interestingly, if that was true, then full fluency in a purely linguistic Turing Test would imply a test of embodiment without incorporating robotics or suchlike. But this Dreyfusian argument seems to be wrong, unless one takes a very 'thin' and somewhat self-serving view of the body, because humans who are close to being disembodied by disabilities can be fully intelligent. Effectively disembodied sociologists can also be intelligent in respect of some targeted culture; they can acquire interactional expertise and understand sub-cultures through immersion in language without practising and without becoming contributory experts. But this ability to work without a body does not apply at the collective level: humans would never have developed the concepts they did, had their bodies not been formed as they are. For example, to be bipedal and have opposing thumbs was *necessary* for the development of the language of bat-and-ball games – that is the bottom-up part of it – but it is not *sufficient* for the development of bat-and-ball games, leave alone ping-pong and cricket in particular. So the body has a bivalent relationship with artificial intelligence.

The bivalence shows itself again in the way that humans learn esoteric knowledge and originate new scientific knowledge in small, closed groups, with boundaries managed by trust. Humans find ways to *limit* contributions to the creation of new knowledge – they do not try to maximize contributions. Because trust is at the heart of these interactions, face-to-face interaction is important and this brings the human body back into focus: humans build trust through eating and drinking together, by smiling, and through body language in general; the right kind of body language can help supply the context to make it possible to say one thing while meaning another and having it understood, an ability that is important to humans; these are examples of some of the polimorphic actions that are central to knowledge creation. The ambivalence is there because not every individual has to share in the bodily interactions; exceptions can be made for those who are unable to take part. Once more, we have to separate the collective activity and the activity of specific individuals. Among humans there is a division of labour, which makes it possible to contribute to science without these body-dependent 'interactive abilities'. But mimicking human activity in general, even to Level III, probably does require some kind of bodily engagement among most of the participants. Once more, the *testing* of interactive abilities can be managed through testing linguistic skills alone because without the interaction the language would be wanting.

The internet and human culture formation

Once we are paying the right kind of attention to the way culture is poured into artificial brains we can ask again whether consuming everything on the internet is the same as absorbing a culture. If we take the paradigm of culture consumption to be the sociologist absorbing a new culture, we have seen that there are marked differences between the two, especially where the culture is involved in the origination of new patterns.

What has been gained with deep learning's ability to draw on all the knowledge on the internet – and, potentially, to draw on much or most human linguistic interaction – is that the problem of archaism has been greatly reduced. Today's and tomorrow's deep-learning computers are, and will be, continually updating what they know; the frontier of computer knowledge is moving closer to 'now'. When we deal with computers we are no longer dealing with frozen moments in time. We are still looking backwards, but only just.

It remains, however, that the origination of new patterns, in the way humans do it, involves more than keeping up with the frontier: it means helping to create it. Computers can undoubtedly create, but legitimate rule-breaking and precedent-setting in society depend on the subtle, multiple, context-recognition with which the argument of this book began. Often, it also turns on the resolution of lively disagreement among humans. It is clear that learning from the internet is not the same as taking part.

The new artificial intelligence and a new relationship

Kurzweil complains about the relationship between the AI-believers and the critics. He says that AI can never win because the critics are always moving the goalposts. Achievements like beating the world chess champion are belittled as soon as they are achieved – 'that's just chess'. Achievements like that of Watson winning at *Jeopardy!*, now coping with puns, double-entendres, and all the things that critics, including me, once said no computer could cope with, are said to be 'just brute force' (as we are saying here). To repeat, the self-driving car, once thought to be impossible, will soon be on our roads but a single crash from unanticipated causes (as recently happened with a Tesla running into a white trailer and killing its driver), will be taken to nullify its achievements. As soon as something once criticized as impossible is achieved, it is written off as not the real thing. Kurzweil illustrates his frustration with a nice joke:

> The complaints [about Siri's occasional mistakes] remind me of the story of the dog who plays chess. To an incredulous questioner, the dog's owner replies, 'Yeah, it's true, he does play chess, but his endgame is weak.' (Kurzweil 2012: 161)

As we have seen, one of Kurzweil's answers to critics who say it is all brute force is that the brain itself is just a statistics-based pattern recognizer, so these brute- force methods *are* the real thing. That seems to me to be the right *kind* of answer irrespective of whether or not it stands up. It probably isn't the right answer – just the right *kind of* answer – because it faces up squarely to the problem.[110] But the other kind of complaint – that every time we accomplish something the critics say cannot be done, they move the goalposts – is misplaced; that is the *wrong kind* of answer, even though it correctly describes what happens.

The reason is that even the ideologically driven AI community should embrace its failures as readily as its successes; if someone moves the goalposts that should be a cause for celebration because it further clarifies the ultimate task. Moving the goalposts should be a joint task, shared by the protagonists and the critics, just as creation and criticism are bedfellows in the making of knowledge in the natural sciences. There is cause for hope because there are important leaders of the AI community who have not only helped to develop the discipline but also pushed forward the understanding of its limitations; Joseph Weizenbaum, Terry Winograd and, more recently, Hector Levesque and Ernest Davis, come to mind.[111] To help push forward the frontiers one must strive to see the glass of AI as half-empty not half-full.

Pagerank

The problem of embedding can be looked at in a less abstract way. With Google, the order in which responses to googled inquiries are returned is controlled by an algorithm called 'Pagerank'. Pagerank lists returned items according to their popularity as represented by the number of links to a page or some such, reflecting what is being talked about in society. Pagerank puts whatever or whoever is using Google in touch with society's current concerns – very human-like! But Pagerank has a couple of problems. The first, which is not deep, is that those with a commercial interest have found ways to 'game' it so as to give their websites more salience. The more serious problem is that popularity is not validity – once more, quantity is not necessarily quality; the difference between having all the information there is, available to instant access, is not the same as having access to the small group of experts that should be trusted; this is becoming evident and everyone, inside and outside AI, is noticing it. Pagerank's meta-expertise is deficient.

Consider that at the time of the mumps, measles and rubella (MMR) vaccine revolt in the UK in the late 1990s, groups or individuals expressing opposition to the vaccine and support for the rebellion's instigator, Andrew Wakefield, would have had a very significant presence on the internet. But the opposition to the MMR and Wakefield's statement were based on no scientific evidence whatsoever and are now frequently associated with Wakefield's financial interest in single-shot vaccines. A crude Pagerank would have reflected the salience of Wakefield and his supporters and stoked the panic. Google, aware of this problem, is making efforts to find another principle for the ranking of papers:

> The quality of web sources has been traditionally evaluated using exogenous signals such as the hyperlink structure of the graph. We propose a new approach that relies on endogenous signals, namely, the correctness of factual information provided by the source. A source that has few false facts is considered to be trustworthy. . . . We call the trustworthiness score we computed Knowledge-Based Trust (KBT). On synthetic data, we show that our method can reliably compute the true trustworthiness levels of the sources. We then apply it to a database of 2.8B facts extracted from the web, and thereby estimate the trustworthiness of 119M webpages. (<http://arxiv.org/pdf/1502.03519v1.pdf> p. 1)

This approach could be valuable in estimating the trustworthiness of domains but is unlikely to help when it comes to competing scientific hypotheses. One would expect nearly all the facts reported on fringe science websites to be correct and all the facts but one in a paper like that written by Joe Weber in 1996, about the correlation between gravitational waves and gamma ray bursts, to be correct. The one incorrect fact (as most would see it) would be the paper's main hypothesis, but since that is the very thing under dispute in the discussion, it cannot be taken as incorrect before the debate. In the case of a paper like that, the community has decided it is incorrect on the basis of what they feel about the social position of Joe Weber, and the decision cannot be fact-checked. This decision to disbelieve because of Joe Weber's position in the field of gravitational waves in 1996 is not registering a fact; it is creating a fact. Joe Weber's claims are no longer trusted, but that is an input to the process of pattern-making in respect of gravitational waves, not the bottom-up detection of a pattern.[112]

Sciences are social enterprises and they work in different ways. The physicists are always leaning over backwards to prove themselves wrong – always refusing to accept that they might have found something if there is the tiniest chance they might be mistaken; they want the truth and nothing but the truth. The more ambitious members of the AI community often seem engaged in trying to win a game against an opposing team – the critics – rather than searching for the truth (with the critics often falling into the same pattern). Yes, the goalposts are always being moved by the critics, but the real problem is that it is the critics who are moving them. If the ambitious members of the AI community seriously want the truth – 'Have we really accomplished Level III/Level IV/passed the Turing Test?' – they should be continually moving the goalposts themselves so as to make the accomplishment as difficult as it can be. To repeat, when AI was an orphan discipline starved of funds there was an excuse – even if not a very good one – to put the best possible gloss on what was being accomplished; nowadays, those at the frontiers of the AI community are pretty well the most powerful body of research scientists in the world, and without their input the policing and evaluation of AI's accomplishments could not be as thorough as it should be. They should, like the physicists, become an object lesson for the proper conduct of science. I know this is possible because I've seen it with the physicists. It would mean always trying to find new ways to show how what has been done is less than it appears, rather than more than it appears; it is the engineering equivalent of Popper's notion of the falsification imperative, a central norm of science.[113]

Kurzweil is right in thinking that linguistic fluency is the key to knowing whether general intelligence has been achieved in machines and that the Turing Tests that are deployed should be the most demanding kind we can invent. But to make the Turing Test demanding enough will take the skills and understanding of experts of all kinds. So the new emperors of artificial intelligence, I am suggesting, should do something that goes against the grain of the marketing success that has made them emperors. They should shut down their conferences where crowds come to whoop at the success of the latest conversational partner, or at least modify them so that every demonstration of conversational success is followed by a demonstration of conversational failure – at least as long as it is still possible to find and demonstrate those failures. There should be a committee, funded and staffed by the emperors, to control conversational pollution in respect of computers, rather as car-makers are supposed to measure and report their cars' exhaust pollution. And, of course, the whole point will be that the makers would not game the public in the way Volkswagen and other car manufacturers did – quite the opposite. The new AI industry is young enough for its founders to be young enough for this still to be imagined.

The Editing Test and Other New Versions of the Turing Test

As I remarked in the first chapter, we have now seen the first sensible, purely language-based, Turing Test competition, an important and refreshing change from the nonsense that has gone before. The new test was based on Winograd schemas.[114] Hector Levesque and his colleagues focus on the challenge set for computers by common-sense knowledge, which is what I would refer to as 'ubiquitous expertise' – the kind of expertise you need to live in society.[115]

Winograd schemas are sentences in which the reference of a pronoun changes when one word is changed. Understanding them rests on knowledge and understanding of the world, which, at first sight, cannot be circumvented by statistical means. The first example was invented by Terry Winograd:

> The city councilmen refused the demonstrators a permit because they [feared/advocated] violence.

The 'they' refers to either the councilmen or the demonstrators, depending on whether the word in the square bracket is feared or advocated. The trick would be to ask, in a Turing Test, what the reference was according to which of the alternative words was inserted. In this case, if the word was 'feared',, the reference would be 'the city councilmen'; if the word was 'advocated', the reference would be 'the demonstrators'.

Another well-known example is:

The trophy doesn't fit into the brown suitcase because it's too [small/large].

Levesque and his colleagues have developed an 'oven-ready' set of Winograd-schema questions that can be incorporated into truly demanding Turing Tests whenever the need arises. They want to formalize the input to Turing Tests and minimize real-time human intervention and eliminate real-time judging by having the answers pre-tested on human respondents, then stored. The test would be automatically marked:

> . . . my position is that common sense is needed when faced with situations that are "new", that is, sufficiently unexpected and unlike anything seen before. The idea with using sentences in a test is that it is not hard to make sentences that are new in this sense, that is, sentences that talk about new situations, even when the words themselves are familiar. So yes, the actual Winograd sentences I would use in a test would have to be new to the subject (and 'Google-proof' as I called it in a paper), and yes, you may still need experts to construct a store of examples (and pretest them to make sure they work as expected etc.). But the hope is that if the tester does this in advance, the actual testing of subjects can still be done in a mostly automated way. (Levesque, personal communication, 18 May 2017)[116]

The 2016 Winograd challenge was run in New York.[117] Sixty multiple-choice questions based on Winograd schemas were set and six computers were volunteered to try to answer them. The average correct number of answers would be 44 per cent if the multiple-choice answers were simply selected by guesswork or reference to a random number table, while humans, as tested by the organizers, get 90 per cent or more correct. The actual success rate of the six computers, when rounded, was 45 per cent, 32 per cent, 48 per cent, 48 per cent, 58 per cent and 58 per cent.[118] The test had a potential second, more difficult stage, but no computer scored well enough on the first stage to reach the second stage.

The result, then, was clear failure for all the programs entered. Disappointingly, however, there were no entries from teams with great resources and technical skills. I asked Yoshua Bengio some questions about the absence of powerful participants from the test (private communication, 5 September 2017):

Collins: Do you think deep learning would pass such a test? [i.e., one based on Winograd schemas]

Bengio: Not right now, for sure. Yes, in the future.

Collins: No serious teams put themselves forward for the test. Do you know why not?

Bengio: Because we already know that our current models don't have enough common sense embedded in them in order to do a good job at this and because there is not enough 'training data' of that exact kind to rely on supervised learning as we currently do for other things.

Collins: And would you or your colleagues consider competing in the next one (2018)?

Bengio: Not until the basic research (of the kind I am doing) has been done to enable learning agents to do unsupervised reinforcement learning in their environment as to capture the causal structure of it, how to control it, how to represent it in order to plan solutions to tasks (this will first be done in simulated videogame-like environments), and then transpose that in a large-scale setting with humans in the loop and dialogue as the interaction medium.

I asked Geoffrey Hinton the same questions but he replied more discursively, explaining that, two or three years back, he had arranged for a colleague, Ilya Sutskever, to try to resolve Winograd schemas in a test of their own. The experiment was a matter of translating the sentences into French when the 'understanding' of the computer would be gauged by whether the pronoun in the translated sentence had the right gender (where the gender is different according to which word is retained). The idea of translation as an indicator of which word in the sentence is being referred to can be found first in Terry Winograd's PhD thesis.[119]

We, and you, dear reader, can try this out with Google Translate; nowadays it is based on deep-learning techniques, which is why it is so much better than it was a short time ago (<https://www.nytimes.com/2016/12/14/magazine/the-great-ai-awakening.html?mcubz=3>). What this means is that when we use Google Translate we are looking at the frontiers, or something close to the frontiers, of AI and deep learning.

We can ask Google to translate the sentence to do with trophies and suitcases into French. (One must write 'it is' not 'it's' or the genderless 'c'est' is returned). The two translations returned (September 2017) were:

Le trophée ne rentre pas dans la valise marron parce qu'il est trop grand.

And

> Le trophée ne rentre pas dans la valise marron parce qu'il
> est trop petit

The first is a correct translation but in the second the pronoun
is male not female, incorrectly referring to trophy not suitcase.[120] Clearly, Google Translate does not understand the
sentences, that is, it does not recognize the implied contexts.

Hinton told me that in the much more extensive test run
by Sutskever, the results were no better than chance. He went
on to explain:

> A machine translation net with a billion parameters only has
> as many parameters as a cubic millimetre of cortex. That's
> one voxel within a brain scan [a voxel is a brain-scan pixel
> with the third dimension of depth added]. I think we will
> need at least a trillion parameters [a thousand times as many]
> to contain all the necessary world knowledge (assuming
> you don't get to see the test sentences until after you have
> [tried to translate] so they could require *any* common-sense
> knowledge). Modelling a cubic centimetre of brain [1,000
> times the volume of a cubic millimetre] is still some way
> beyond what we can do.

We understand, then, why the powerful players did not
expose themselves to the Winograd schema challenge – they
knew it would be pointless and will remain pointless for some
time to come. We see that it is far too early to consider that
deep learning has solved the problem of normal fluent conversation and any optimism has to be based, not on results,
but on the assumption that progress will continue at a fast
rate, which it probably will: a factor of 1,000 is not much in
the context of Extended Moore's Law. So *if Hinton is right* we

should see Google Translate handling Winograd schemas in the not too distant future. Remember, Kurzweil predicts that the Turing Test will at last be passed in 2029, and presumably he has Hinton's kind of calculation mind. I don't think so, but we shall see – exciting, isn't it!

Another approach that turns on common sense is provided by the simple but ingenious questions invented by Ernest Davis.[121] Here are three out of a large number of examples:

> Mary owns a canary named Paul. Does Paul have any ancestors who were alive in the year 1750?

> Is it possible to fold a watermelon?

> George accidentally poured a little bleach into his milk. Is it OK for him to drink the milk, if he's careful not to swallow any of the bleach?[122]

The intriguing thing about these questions is that because they are so commonsensical no one teaches, talks or writes about their solutions, though we all know what the solutions are as soon as we think about them; the solutions are immanent and 'known' only tacitly until circumstances like this cause them to be explicated. For this reason it is unlikely that the answers can be found on the net. Thus, if, on 28 August 2017, I google 'Can I fold a watermelon?' I won't get the answer I want. Once more, the idea is to pre-prepare a test with multiple-choice answers that can be pre-tested on humans and marked automatically without the use of real-time expertise.

At first sight I could not think of any way for computers to learn to answer the questions; I don't even know how I know how I answer the questions – they are elements in my body of 'relational tacit knowledge'.[123] So it seems that this, together with the Winograd schemas, provides an excellent way forward for the first extension of the Turing Test. But given Extended Moore's Law they don't seem future-proof. But before getting on to this, let us examine one other aspect of the approach of Levesque, Davis and colleagues.

Levesque draws attention to the so-called super-human fallacy, described by Papert and Minsky in the early, antagonistic days of artificial intelligence.[124] Adapting Levesque's version of the fallacy, what follows is that the standard demanded of a computer which is to be counted as passing the Turing Test cannot be equal to the very best of human performance because hardly any humans can achieve such a standard; we cannot ask the computer to compose a work of Shakespeare or a brilliant poem.[125] In Chapter 3, I skate over the problem by saying we need a standard for resolving text correction that would be achieved by the top 0.1 per cent of copy-editors. But it may be possible to explore the problem more systematically in terms of the 'three-dimensional model of expertise'.[126]

The problem with many traditional philosophical and psychological analyses of expertise is that they start by assuming that expertise must be something esoteric. In the earliest days of AI, this confusion could be what led to the assumption that language-handling would not be a crucial problem – if pretty well everyone could handle language, it shouldn't be a serious obstacle for computers that could do lightning calculations beyond the capacity of any human. The later emphasis on the difficulty of common-sense knowledge can be seen as an attempt to put the matter right. The three-dimensional model starts from the assumption that acquiring expertise is mostly a matter of acquiring the tacit knowledge pertaining to a domain. This would be 'somatic tacit knowledge' in the case of something like playing a violin or balancing on a bike, while relational tacit knowledge would include

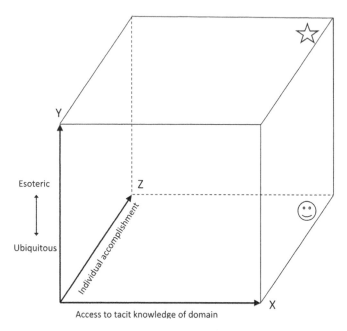

Figure 10.1 *Three-dimensional model of expertise*

common-sense understanding of the physical world, and it would be 'collective tacit knowledge' in the case of fluent language-speaking. Of the three dimensions in the diagram, the first is the traditional dimension of personal capacity – some people are just better at doing certain things than others – and this is the Z-dimension, going into the page in Figure 10.1 and labelled 'individual accomplishment'.

The X-axis – the horizontal axis – represents the extent to which an individual immerses themselves in the society of those who possess the tacit knowledge: the length and depth of the apprenticeship or, given the importance of interactional expertise, the extent of linguistic socialization. The vertical Y-axis indicates the extent to which the expertise is a widespread ubiquitous expertise, such as native language-speaking, or an esoteric expertise such as gravitational-wave physics.

I think there could be some confusion between the dimensions in the debate about what kind of person a computer should represent in a good Turing Test. I think what Levesque wants to argue is that the computer should not be tested for its ability in respect of an esoteric expertise, such as being a first-class playwright or poet, and I agree with this. To conduct a proper Turing Test with an esoteric expertise as the target would require gathering together humans with similar rare and exotic expertises to act as the 'controls' and judges; this is not what we want to be doing (except in cases such as my trying to pass as a gravitational-wave physicist). So we can all agree that for the kind of future Turing Tests we have in mind we should be looking at ubiquitous expertises, such as native language-speaking.[127] Thus, we will be looking towards the bottom of the 'expertise space diagram' (which is another way of referring to Figure 10.1). But surely we still want to go a long way along the X- and the Z-axes to find our target for the Turing Test – when it comes to ubiquitous expertises we want to compare the computer with someone who has maximal *savoir faire*; the possession of *savoir faire* is one way to describe a virtuoso in ubiquitous skills. When it comes to language and the X-axis, we don't want to compare computers with young children who have not yet fully acquired natural language-speaking skills, but want to compare them with adults who are as fluent as possible, barring a degree of fluency that transports them up the Y-axis into the esoteric realm of playwrights and poets. And, on the Z-axis, we don't want to compare them with autistic persons or others with some kind of condition that reduces their ability to understand all the nuances of their native language – we want to compare them with the best native-language speakers we can find. If we don't go far enough along the X- and Z-axes we will be making the task too easy for computers and we will not be exploring the full range of human abilities. This seems to me to justify something like the top 0.1 per cent of copy-editors being selected as the 'target expertise' in the Turing Test. In terms of the expertise space diagram, we should be looking to reproduce the abilities of persons in the bottom, back, right-hand corner; the approximate location of the smiley face in Figure 10.1. In an ideal world those who design the tests would, however, need to know the esoteric intricacies of how the programs work so as to exploit their every weakness, and they should come from somewhere around the *top*, back, right corner near the location of the star.

In passing, let us note that the super-human fallacy seems to be no obstacle for AI enthusiasts when we are talking of tasks such as winning at chess or Go. In such cases the enthusiasts loudly proclaim their mechanical prodigy's success at beating every possible human. Why don't we expect the same of computers when it comes to natural language-speaking? The answer surely goes back to the distinction between tasks with technically defined and culturally defined goals (Chapter 5). You can't beat every human at language-speaking because fluent language-speaking is not the property of individuals; it is something that individuals borrow from the society in which they are embedded, so for one human to try to beat every other human at fluent language-speaking makes no sense. What counts as fluent is continually being redefined by the collectivity of humans, so the logic of beating every human would be something like the logic of trying to be more average in weight than any other human! I think this illuminates the difference between technical and culturally defined goals a little further and shows, once more, just why embedding in society is a logical necessity of intelligence as measured by a proper Turing Test.

The editing test and its advantages

Going back to Davis, Levesque and others, I do not think any pre-prepared test is going to be future-proof. Indeed, it has been pointed out to me that at least some of the questions are scarcely present-proof since the use of 'word embedding' indicates a way forward.[128] A corpus of words is available that can be used to investigate the relationship of words/concepts in English as it was in 2007 (at: <http://bionlp-www.utu.fi/ wv_demo/>). If I use that website as of now to investigate the watermelon-folding problem, I find that the relationship of 'watermelon' and 'fold' in the corpus is represented by the number '0.03378', whereas in the case of 'watermelon' and 'cut', the number is '0.284858'.[129] Another website enables us to list the fifty words that occur most frequently within three words either side of 'watermelon'; they are:

> punch, man, watermelon, white, eat, seeds, tourmaline, seed, strawberry, eating, rind, seedless, slice, cantaloupe, pink, red, sugar, company, juice, fruit, festival, chicken, wild, tomatoes, salad, slices, sweet, pineapple, ice, mosaic, tomato, melon, juicy, flavor, crawl, vanilla, fresh, audio, spitting, electric, pickles, slim, strawberries, blog, blue, green, woman, wheat, honeydew, pepper. (<https://corpora.linguistik.uni-erlangen.de/cgi-bin/demos/Web1T5/ Web1T5_freq.perl>)

Fold does not appear! This is trivial in itself but shows the potential for using the relationship of words in the language to indicate the relationship between concepts.[130] Maybe word-embedding is interactional expertise in statistical practice . . . The power of the idea is that no one has to be told anything: the relationships one is uncovering arise out of a 'silence' in the language, just as was claimed above (p. 162). The silence, in this case, is the absence of the word 'fold' from the region of the word 'watermelon' in the spoken language. Something of practical importance can be learned here by the fluent language-speaker without ever being told about it; the transmission of knowledge of practical importance is being effected tacitly! The knowledge is in the language without its being explicated, just as is the claim for interactional expertise.[131]

If we allow ourselves to imagine some of the futures that might arise out of exponential growth of Extended Moore's Law, we can see other possibilities. Thus, consider Winograd schemas: as it is, one can find lists of such sentences on the internet. One list with 177 entries, constructed and collected by Ernest Davis himself, has correct answers in the following form:

1. The city councilmen refused the demonstrators a permit because **they** feared violence.
   Snippet: **they** feared violence
   A. The city councilmen
   B. The demonstrators
   Correct Answer: A
   Source: (Winograd 1972)
2. The city councilmen refused the demonstrators a permit because **they** advocated violence.
   Snippet: **they** advocated violence
   A. The city councilmen
   B. The demonstrators
   Correct Answer: B
   Source: (Winograd 1972)[132]

So 177 possibilities are eliminated, assuming that deep-learning computers are, or will be, so well integrated into the net that they can learn from it.[133] Somehow, new questions of both kinds will have to be generated and stored if they are to be pre-prepared for use in the Turing Test, but the process has to be done in such a way that it is proof against the scenario where computers are 'watching' everything we type on our computers or smartphones and listening in to everything we say on the telephone. And, remembering HAL from *2001*, we have to anticipate that if we are in view of an internet-linked camera, anything we write will be deciphered and anything we say will be entered into the database via lipreading! So when these new examples are being worked out, humans will have to stick to paper and pencil and work in silence in closed rooms without cameras, while the new list of tests will have to be stored in a bank vault or equivalent. This all seems a bit like science fiction, but the implications of an Extended Moore's Law are like science fiction – we struggle to imagine where exponential change will take us. The new kinds of test work brilliantly today – even too brilliantly as the unwillingness of powerful players to participate in the test in New York demonstrates – and future annual tests will, no doubt, continue this trend. But what of the longer term?

It could also be that much of the common-sense knowledge needed to resolve Winograd schemas and Davis's questions could be stored in formal ways in computers equipped with databases developed by projects such as Lenat's CYC.[134] CYC aims to strip and store all the knowledge in all the encyclopedias in the world and, no doubt, any lower-level common-sense knowledge stored in other reference works. I am told that CYC also employed armies of graduate students and the like to try to formalize common-sense knowledge. Perhaps CYC's knowledge could be combined with deep learning to answer even newly invented and secret Winograd schemas, or knowledge questions, because these schemas still rely on existing knowledge about the world. Even all the things you are never taught, as in the Winograd schemas and Davis's ingenious questions, could perhaps be programmed

in by the armies of human helpers continuing to formalize common sense.[135] Relational tacit knowledge can be
explicated – 'for example, you can only fold things that have
one dimension much shorter than the other two; melons have
three similar dimensions' – and even though it cannot all be
explained at once, enough of it might be explicated by some
kind of future means to allow computers that do not have
full human capacities to pass the test. I must admit that the
'amount' of common sense seems so large that I have trouble
imagining a CYC-type solution, but one has learned that it is
hard to overestimate the power of this kind of solution so it
seems better to err on the safe side.[136]

For this reason I am going to anticipate the next move of
the goalposts should we ever reach a time when Winograd
schemas and common-sense knowledge can be handled by
AI. For the most revealing kind of confrontation as the future
unfolds, I suggest that we look towards testing computers
on the moment-by-moment newly created knowledge that
is associated with real-time rule-breaking, precedent-setting
and repair in conversation; we can be sure it is impossible
to foresee every way of misspelling or inventive handling of
words because the number of combinations soon exhausts the
universe.

I am proposing that an element in the right kind of demanding Turing Test would be a real-time 'editing test'. The
editing test also minimizes the amount of pre-preparation and
ingenuity required. The interrogator in an editing test would
insert some non-standard phrases, involving new examples of
rule-breaking and precedent-setting. The editing test would
also check whether the computer had made the transition
from Level II of artificial intelligence, asymmetric prostheses,
to Level III – prostheses that are symmetrical. Symmetrical
prostheses would be as capable as humans at repairing broken
language. The point is that this is the equivalent of engaging

in the same kind of dialogue that led, eventually, to the discovery of gravitational waves, or any similar piece of creative science. But in the case of the way we use words we can make such discoveries every day or even every few minutes; new common-sense relationships in the physical world cannot be invented in that way – the physical world is relatively fixed – the world of language is always in flux. Yes, we can ask previously unimagined questions in respect of the physical world, but because of its relative fixedness and potential explicability there is, perhaps, a remote possibility that all, or nearly all, such questions could be anticipated. On the other hand, because of our native linguistic socialization and consequent repair and editing skills, we can manage constant acts of linguistic creativity, understanding them and sorting the legitimate from simple mistakes, as fast as anyone can create them – and that is fast. Furthermore, anyone with linguistic *savoir faire* can do it in real time. 'I am going to do something funny with a word somewhere in this sentence to <u>ees</u> if the computer knows what to do': there – I've just invented a new version in real time as I type! (The editor/computer, of course, has to change nothing.) And, if I want to be doubly certain about what is going on: 'I am going to misspell "weird" first and then not misspell <u>"wierd"</u> to see if the computer is as clever as my copy-editor.' (Nothing should change in the context of this book, but both weird and <u>wierd</u> should change in the context of a question in a Turing Test. But it needs a lot of thought to understand this; it needs solving by someone well along the X- and Z-axes of Figure 10.1).[137] The claim is that so long as computers are not as deeply embedded in language as fully socialized humans, they will fail at the editing task when compared with top human editors. If they could do it, on the other hand, we would have to say that they have managed to become linguistically socialized and have, by that fact, acquired general intelligence.

To reiterate a point made earlier, not all humans would be able to edit the difficult passages; weak or careless editors would fail. But would *any* computer be able to edit such sentences properly if they had never been encountered before? They require not only contextual understanding of the text but, in one case, contextual understanding of the task – for example, book editing or taking part in a Turing Test. The big claims about artificial intelligence are, as remarked earlier, binary: if it is claimed that Level III or Level IV of AI is being accomplished, then it must be shown that there is a computer that can accomplish at least what 99.9 per cent of human editors can accomplish. Once more, to make really good tests of this kind, both left-hand and right-hand kinds of expert need to be on the same side, jointly inventing tests that are sufficiently novel and demanding.

This kind of activity is important, first to avoid the Surrender, but also because the aim of the most ambitious, and certainly the most ideological, wings of AI, is, finally, to destroy the special claim to uniqueness for humanity and to show that we are merely machines. Whether there is something special about us is, perhaps, the most important question the human race has ever addressed. If we are to believe the answer, all parties will have to use every ounce of brilliance and ingenuity at their disposal to search for the remaining indications of uniqueness and to offer understanding of just what new and, for most of us, unforeseeable, brilliance in the way of brute strength and other 'tricks' might have been deployed to seem to resolve the problem when it has not really been resolved at all.

We have already seen that expert analysis is needed to answer some of the relatively easy questions about deep learning. What exactly is the difference when 'unsupervised' pattern recognition of animals, motorbikes and violins and the like is tried with a selection of completely random photographs, as compared with photographs where subjects are centred? How much labelling, explicit or implicit, is going on during the training? Maybe experiments along these lines have been tried, maybe not: experts from the left-hand side should know or, if they don't, they have the resources to try them.

The outcome of these experiments and the properly conducted Turing Tests should be promulgated to the public, who ought to be encouraged to examine them with more fingernail-biting anxiety than they give to supporting a sports team or a pop star. If I'm still around I'll be biting my nails when the powerful contenders consider it time to subject their programs to sensible Turing Tests and, should they surmount the hurdle first time, I hope this book will encourage the next step, making the next competition more difficult, and still more exciting, and so on. What a prospect! So long as the computers continue to fail the increasingly difficult tests, citizens will be discouraged from adopting a lazy anthropomorphism even when they are surrounded by self-driving cars, robot carers and near-infallible automated medical diagnosticians – thus will the Surrender will be delayed a little longer. In the meantime, we are all, or nearly all, expert language users and we can, and should, be continually testing for the weaknesses and deficiencies of language use by computers. We can do this by using our own disenchantment devices: run the appropriate Winograd schemas though Google Translate – look for ones taken from Davis's list,[138] with pronoun references that are differently gendered in French, or invent new ones! Or try out the editing test with my examples or new ones of your own. Or easier still, since your smartphone is linked to some of the best artificially intelligent computers in the world, give Siri and its ever improving equivalents some tricky transcription tasks, such as speaking into it: 'Siri, I want you to misspell weird in this sentence with the "i" before the "e"' or some such, or try mumbling normal sentences badly or in a loud-noise background. Transcription tasks could make for some slick and easy full-scale Turing Tests too; all these tests can be tried on humans for comparison both by you and by a Turing Test committee. Invent your own disenchantment devices and keep trying them whenever the new breakthroughs in artificial intelligence are announced; try it in the pub and the café with friends. At the time of writing this won't be very exciting because the spelling and grammar checkers, and the transcribers, will simply fail. But the 'automated editor in the sky' will be getting better and better, so the tests might become a little more worthwhile as time goes on. Remember, we should do this because it is interesting, but more importantly, because it is 'the use of words and general educated opinion' that will organize, both for us and the machines, our place in the world.

Appendix 1

How the Internet Works Today

Cmabrigde Uinervtisy

If I type the misspelled passage about Cmabrigde Uinervtisy into Google's search field it will respond. The first URL that is returned is: <https://www.learnthat.org/news/cna-yuo-raed-tihs/>. It is a story about the 'ten-year-old' Cambridge experiment.

Somewhat more surprising is the result if you insert the senseless passage with words jumbled into the search field:

> at aoccdrnig be olny rsceareh a it mtaetr in oerdr the waht dseno't to ltteres a are, the iproamtnt pclae Uinervtisy, taht tihng lsat is the and wrod Cmabrigde ltteer in the in rghit frsit.

Google returns a similar list of URLs having, I presume, recognized individual garbled words as belonging to the orig-inal Cambridge experiment. So here Google is doing things differently from a human since it seems pretty indifferent as to whether the passage makes sense or not, whereas we humans would not be. But my book, *Tacit and Explicit Knowledge*, where I compare longer versions of the extract from the Cambridge experiment and my senseless, disordered, version of it, comes up on the first page in both searches, and it turns out that Google hasn't deciphered the disordered-words passage after all, but picked the precedent from my book.

Now let us try something different. Let us invent a new garbled passage. This is a garbled but sense-making version of a letter to my doctor – something that has not been and will not be published anywhere, at least not by me.

> I am rlaley srroy auobt msniisg my atnpipeonmt wtih you bekood for 8.40am on 13 Jnue. It is no eucsxe but, as yl'uol see form yuor rrcdoes, taht day is my bhadtriy and wehn I akwoe my wfie gvae me my crad and a nebmur of pertsnes and we set auobt pnnainlg the day – the anpnpteoimt wnet rhgit out of my haed

I think you, reader, should be able to read it pretty easily in the way you could read the unjumbled 'Cmabrigde Uinervtisy' passage.

Putting that into Google's search field on 5 July 2016 produced a no-return – it 'did not match any documents'. Somewhere down the line, however, if this book gets picked up in some web-accessible form, that passage will come to be recognized by Google. Try it, reader – if Google does not recognize that this passage comes from this book, then we can pass over the next couple of paragraphs – perhaps coming back to them in a year or two when the book might have been picked up. If Google does recognize the passage as belonging to this book then we can go on and try a couple more experiments.

The next thing will be for you to 'translate' that passage into normal English. I am not going to do it here because if I do, the translation will be published in this book, and I don't want the translation to be published. But if you, reader, write out the ungarbled version and type it, or cut-and-paste it into Google's search field, I am betting you will obtain similar results to mine: when I did that, on 5 July 2015, Google returned a list of haphazard entries that did not bear upon the passage in any close way. The obvious moral is that Google isn't interested in sense-making. It will be able to find a suitable response to a garbled passage provided it has been published, but it won't be able to find the right one for a passage when it is ungarbled into proper English so that it makes complete sense, if that has not been published. Of course, somewhere down the line anyone could sabotage the whole experiment by putting all the passages, including the sense-making one, on some website.

If you want to try more experiments that will sidestep this kind of sabotage you can construct some garbled passages by taking any piece of your own prose and keeping the first and last letters of each word the same, while rearranging the letters in the middle of the word. This should produce text that is easily readable. You can then try jumbling the word order and prove to yourself, or to someone who is not already familiar with that passage and its transforma-tions, that it becomes unreadable. And, of course, you can try out what happens when these passages are offered to Google.[139]

What we are illustrating here is that Google and the inter-net are always looking backwards, though the frontiers of what they look back to are always being updated to some-where nearer today. That is how the new AI differs from the old AI where the frontiers didn't move. The same point is illustrated by 'little dogs': they are close to invisible on the internet (at the time of writing) because they are a term that belongs to the craft of gravitational-wave detection, a term known only to the inner circle (my books aside).

Appendix 2
Little Dogs

Detection of gravitational waves requires a coincident burst of energy on at least two widely separated detectors. The first discovery involved detectors in Louisiana and Washington State separated by about 2,000 miles. But to make such a discovery convincing, it has to be shown that such a thing could not happen by chance and the problem is that the detectors are always beset by 'glitches' – bursts of energy that are due to noise not signal. To estimate the chance of a couple of loud coincident glitches being produced by noise alone, a series of 'time-slides' is constructed: the output from one detector is 'slid' along that of the other detector in time-steps and a search for coincidences is made at every time-step. Such time-offset co-occurrences of energetic bursts could only be caused by noise alone since there could be no properly coincident cause for them. The statistical confidence is calculated from the number of such offset 'coincidences' of similar energy to the genuine coincidence in many millions of time-slides.

The problem discussed here occurred during the search

for the blind injection known as Big Dog – named for the putative source, Canis Major (which gives rise to the canine nomenclature). A little dog arises when, in a time-slide, the excursion belonging to the putative coincident signal in one of the detectors occurs opposite an excursion in the other detector which is unequivocally caused by noise. If the excursion that is a putative signal does turn out to be a real signal, then this little dog is not a proper offset coincidence since one half of it is not noise at all but a genuine signal. Therefore, it should not feature as an offset signal in the probability calculation. But if the putative signal turns out not to be a real signal, then the little dog *is* a proper offset signal and should feature in the noise calculation, reducing the statistical confidence in the putative signal. The trouble is that we only know whether or not the putative signal is a real signal after we have done the noise calculation, so the calculation of the statistical confidence contains a circularity. This is the source of the argument about whether little dogs should be included in noise calculations; I try to argue on a priori grounds, in Collins (2013a), that they should not.

Notes

*Chapter 1 Computers in Social Life and the
Danger of the Surrender*

1 The Turing Test will be discussed at length in Chapter
3. For those who are not familiar with it, the 'stripped
down' description involves a 'judge' asking questions,
via a keyboard, of a hidden computer and a hidden
human. If the judge can't tell which is the computer
and which is the human, then the computer is said to be
intelligent.

2 The idea was invented in the 1960s but made salient
in the 1990s by mathematician Vernon Vinge, who
said the disaster would come about between 2005 and
2030. Ray Kurzweil, one of the most forthright of the
contemporary seers, puts the date at 2045. Incidentally,
there are many ways that computers could destroy us
without being super-intelligent. Nowadays they control
so much that we can imagine bugs in the programs – and
bugs can never be fully eliminated – leading to all the

world's nuclear weapons being launched, or some such, or we can imagine malicious human hackers engineering the same thing. Or maybe when machines are taught to reproduce they will run amok exponentially and use up all the world's resources on making copies of themselves. These possibilities or some partial versions of them are more worrying and immediate than the Singularity.

3 Since the flag is not preserved when the computer prints out a document I had to add underlining. In the book we see a double underline in place of the original, red, jagged underline that appears in my manuscript.

4 Hector Levesque (private communication, 22 May 2017) points out that in good English the 'wierd' would have quotation marks around it as in this note. But my copy-editor and you, dear reader, can cope without the quotation marks so it is still a good test for a computer mimicking human abilities. The difference between Levesque and me is that he wants to formalize the input to the Turing Test so that the questions can be put in place in advance, whereas I think that the conversation will have to be created in real time (see Chapter 10).

5 Except in the discussion of Level IV of artificial intelligence – see Chapter 5.

6 See also Alan Blackwell: 'It appears that [Machine Learning] systems are interacting with their social context, for example through the use of big data collected from social networks . . .' (Blackwell 2015: 2 – the stress should be on 'appears').

7 For example, see Table 1.1 on p. 15 of *Why Democracies Need Science*; the mistake might be corrected if there is more than one printing. The neologism is first found in Collins and Kusch (1998). *Why Democracies Need Science* is Collins and Evans (2017).

8 'What if the human and the computer cannot be distin-

guished because the human has become too much like a computer?' (Blackwell 2015: 2).

9 See also O'Neil (2017) for the contemporary attack on humans by computers. And see Ben Schneiderman's Turing Lecture: 'Algorithmic Accountability' (<https://www.youtube.com/watch?v=UWuDgY8aHmU>). Levesque (2017) also sees the problem: 'What I would be most concerned about is the possibility of computer systems that are less than fully intelligent, but are none the less considered to be intelligent enough to control machines and make decisions on their own' (p. 139). The particular contribution I am trying to make here to the problem, that is already well understood by at least some others, is about language understanding and context.

10 There are classic books that take a sceptical look at the 'hype' that has often accompanied fairly modest success in building intelligent computers. Two classics are Weizenbaum's 1976 book and Dreyfus's 1972 volume. I authored or co-authored two such books in the 1990s, *Artificial Experts* and *The Shape of Actions*, which have had modest success. Both centre on the problem of social context. The difference with this book is that it addresses the extraordinary progress – unforeseen when these other books were written – that has been made in recent decades and which makes some of the older critiques outdated.

11 Levesque (2017) calls it Adaptive Machine Learning (AML).

12 In the words of my book *Changing Order: Replication and Induction in Scientific Practice* (1985/92), it is 'changing the order' of human life in just the same way as the discovery of gravitational waves changed the order of human life – see *Gravity's Kiss* (Collins 2017).

13 Interestingly, my computer flagged 'rite' as a grammar

error rather than a spelling error given that rite is correctly spelled in the *context* of religion which is the wrong
context here.

14 For a discussion of the asymmetry point in the context
of the 'extended mind' thesis, see Collins, Clark and
Shrager (2008). For social scientists, 'Actor Network
Theory' gives rise to deep confusion here.

15 That this model of the human is *not* inevitable is also
noted by Levesque (2017: 141).

16 'Now we cause each of the survivors to multiply themselves until they reach the same number of solution
creatures. This is done by simulating sexual reproduction: In other words, we create new offspring where each
new creature draws one part of its genetic code from one
parent and another part from a second parent. Usually
no distinction is made between male or female organ-
isms; its sufficient to generate an offspring from any two
arbitrary parents . . . This is, perhaps, not as interesting
as sexual reproduction in the natural world, but the relevant point here is having two parents' (Kurzweil 2012:
148).

17 (Levesque 2017: 131).

18 Except for Ava, in *Ex Machina*, who is said to have
learned to speak from eavesdropping on the world's
mobile phone conversations; we discuss just this possibility later!

Chapter 2 Expertise and Writing about AI:
Some Reflections on the Project

19 Table 2.1 is adapted from Table 5 in Collins (2010).
Barrow (1999) has also attempted to study the meaning
of the impossible.

20 Here, surprisingly, I seem to be on the same side as
Geoffrey Hinton, the pioneer of deep-learning tech-

niques. As we'll mention again later in the book, he has recently said that AI needs to start all over again: <https://www.axios.com/ai-pioneer-advocates-starting-over-2485537027.html>.

21 For speculation by a physicist about the power and dangers of artificial intelligence, see Tegmark (2017). Tegmark was one of the signatories of the 'Singularity letter'.

22 One can, of course, believe that humans are essentially machines without cleaving to free-market capitalism. What I am suggesting here, admittedly based on little more than what I see seeping into academic discourse (the high point may have passed with the collapsing of the academic respectability of mainstream economics), is the difficulty of getting any argument based on the collective nature of human action accepted among various groups of otherwise excellent thinkers.

23 See: <http://commonsensereasoning.org/winograd.html>.

24 A starting point is Collins and Evans (2007).

25 See also Collins (2014).

26 See the discussion of the 'Locus of Legitimate Interpretation' (LLI) in different cultural enterprises in Collins and Evans (2007: 119–21). For the case of architecture, see Collins et al. (2016).

27 One can see that this 'right-to-left' expertise can work by looking to books written by non-technical experts, which have had an impact on AI. These include the Flores side of Winograd and Flores's (1986) *Understanding Computers and Cognition*, Dreyfus's (1972) *What Computers Can't Do*, and Suchman's (1987) *Plans and Situated Actions*. I also use myself as an example since, to be frank, I need all the recognition from the left-hand side that I can get since my 'left-side' expertise consists of no more than some

self-taught elementary programming in simple computer languages and writing a 'toy' program in the AI-language PROLOG, as reported in my 1990 book. Nevertheless, in 1985, my first paper on the topic of artificial intelligence (Collins, Green and Draper 1985) shared the prize for technical merit at the British Computer Society conference 'Expert Systems 85', held at Warwick University. Subsequently I wrote a book about AI (Collins 1990), which gets a mention in Margaret Boden's (2008) history of AI, and co-authored a second book (Collins and Kusch 1998). According to Google Scholar, as of July 2016, these books together have been cited more than 800 times. They are also favourites with at least some software testers – the professionals who actually have to deal with human–computer interaction (<http://www.developsense.com/blog/2014/03/harry-collins-motive-for-distinctions/>). While the examples I have given refer to works of non-technical experts who directly address AI, many other non-technical experts have had an influence. Ernest Davis told me that 'off the top of his head' the non-AI specialists that he considers to have had a large impact on the field are David Marr, Saul Kripke, John Nash, Noam Chomsky, Charles Fillmore, George Zipf and Michael Bratman.

28 Taking Kurzweil as an example of the left-hand side experts, he does not cite any of what I would consider to be the important past critics of AI, such as Dreyfus; this gives a sense of how some members of the sides represented in Table 2.2 inhabit separate 'silos'. The mathematicians and physicists who consider themselves experts on AI tend to know nothing of these debates.

29 I do occasionally tell physicists what to do in the way of physics, but only as a kind of slightly 'tongue-in-cheek' sociological experiment to see what happens.

30 My previous books on AI, written in the 1990s, were concerned with symbolic AI and expert systems in particular. Some critics from the left-hand side have also chosen to concentrate on deep learning in their critical writings, including Blackwell (2015) and Levesque (2017); this helps affirm my decision not to engage in a much wider survey of AI approaches.

Chapter 3 Language and 'Repair'
31 Blackwell (2017: 11). Alan F. Blackwell is professor at the Computer Laboratory, University of Cambridge. A similar quote is found in his 2015 paper.
32 See Collins (2010: 115) for complete versions of these passages. Later on we will discover more about this example.
33 See Collins and Pinch (2005: Ch. 2), for bogus doctors, or Collins and Evans (2007: 54) for a brief mention of these and other confidence tricksters.
34 Weizenbaum (1976).
35 For a biologist making a similar point but with the power of the brain in mind, see Laland (2017) and for a similar point from an anthropologist, see Deacon (1997).
36 On the topic of non-human mistakes, as one of my earlier books explains (Collins 1990), the pocket calculator doesn't do human-like arithmetic because it does not know how to approximate in context.
37 Readers who want to pursue the original debate can find as much as they want about the Chinese Room argument by looking it up on Google or the like.
38 A more complex design with the same inspiration is due to Ned Block and is discussed in Collins (1990).
39 See Collins (1990) for the argument. A technically proficient reader tells me that the problem is solvable by computers by using partial matches for misspellings

rather than exact matches, so my claim that the look-up table will have to expand so enormously is wrong. But if we allow things like the two 'Cmabrigde' passages – remember, all these misspellings and manglings have to be anticipated – partial matches won't solve the problem of sheer size of the left-hand side.

40 Another example of the problem can be found here: <http://www.theglobeandmail.com/technology/tech-news/how-microsofts-friendly-robot-turned-into-a-rac ist-jerk-in-less-than-24-hours/article29379054/>, the title being self-explanatory. It concerns a system picking up its way of speaking from the internet. The general pattern of these arguments (though bad language is not discussed) can be found in Collins (1990: Ch. 14).

41 See Collins et al. (forthcoming) and Collins and Evans (2014), for development of the Imitation Game as a sociological method.

42 For one of the latest unhelpful claims, see: <http://www. bbc.co.uk/news/technology-27762088>.

43 This remark can be found in his 2005 (n.30 on p. 505) and also in his 2002 essay (<http://www.kurzweilai.net/ a-wager-on-the-turing-test-why-i-think-i-will-win>). This essay is a very good short introduction to Kurzweil's ideas and contains much more sense about the Turing Test and computers than many other sources, though, as will be argued in this book, his arguments are missing something.

44 A number of AI specialists pointed out to me that Kurzweil is considered a marginal figure in terms of recent contributions to the research front of AI, but it was agreed that his ideas are useful for this book.

45 I report statements and conversations at a workshop on deep learning held in New York in summer 2017.

46 Laland (2017); Deacon (1997).

Chapter 4 Humans, Social Contexts and Bodies

47 Some time after writing this passage I discovered that the scenario is reproduced in all details of principle in the film *Arrival*. Unfortunately, the film does not deal seriously with the language problem, resolving it in a kind of magical way with a leavening of mathematics.

48 Note that this *can be done* in determinist environments like the game of Go, of which more below.

49 See Collins (forthcoming), *Forms of Life: The Method and Meaning of Sociology*, for an exploration of this method.

50 Imitation Game is capitalized where it refers to the technical method we have developed for research purposes (Collins and Evans 2014; Collins et al. forthcoming, *Imitation Games: A New Method for Investigating Societies*).

51 Mathematical questions were banned but this is not as serious a fault as it might seem – see Collins (2007). Collins's books on the sociology of gravitational-wave physics are Collins 1985/92, 2004a, 2011a, 2013a, 2017. The first gravitational-wave (GW) Imitation Game was written up as a news item in *Nature* (Giles 2006; see also Collins and Evans 2014), though *Nature* compared it to a hoax, whereas it was a demonstration of genuine understanding. The second, and still more revealing experiment, is described in Chapter 14 of Collins (2017; also at <http://arxiv.org/abs/1607.07373>).

52 I think the idea of interactional expertise confounds Levesque's (2017: 85) division between 'book smarts' and 'street smarts' unless 'book' is construed literally as 'text' as opposed to talk.

53 For the latest discussion of the origins and range of applications of the concept of interactional expertise, see Collins and Evans (2015; a pre-print is available at <http://arxiv.org/abs/1611.04423>).

54 The basic idea is found in Collins (1990), where it is

referred to as the distinction between 'machine-like actions' and 'regular actions'. Philosopher Martin Kusch and I worked it out much more carefully in our 1998 book, *The Shape of Actions*, changing to the more descriptive labelling along the way.

55 To divide actions in this way requires that a rather complicated series of in-between cases be dealt with. The full analysis is found in Collins and Kusch (1998) but it is not necessary to go through the complexities to understand the basic point. We will however need to refer to the in-between category, 'disjunctive mimeomorphic actions', when we get to Go in the next chapter.

56 Private communication, 2 September 2017.

57 Collins and Kusch (1998) is full of examples of the different kinds of actions.

58 See Laland (2017) and Deacon (1997) for other approaches to the centrality of language.

59 I heard this from a very authoritative source who would certainly know the facts of the matter but I have not done any 'detective work' to follow it up, though see his obituary at <http://dailynous.com/2017/04/24/hubert-dreyfus-1930-2017/> – he died in spring 2017. The flavour of academic life at the time can be sensed from, for example, Papert's (1968) critique of Dreyfus.

60 All this is explained at greater length in a number of other places. An accessible starting point is Collins (2014).

61 I chose Wikipedia as it can provide a concise account suited for a wide readership so long as its reliability can be vouched for.

62 A key source for Dreyfus is the book *Body and World* by Samuel Todes (2001), which is based on a PhD thesis written by Todes when he and Dreyfus were graduate students at Harvard. Todes discusses the importance to

our perceptions of our bodies having a front and a back and living in a world where up and down is defined by gravity. The book was originally written as a PhD thesis in 1963, and Todes is solely concerned with the role of our bodies in our experience of the world, and thus writes in his introduction:

> The reader is forewarned that the analyses presented in this study are not of our normal experience in its full complexity. . . . Thus, for example, for the purposes of this study of the human body as the *material* subject of the world, our experience is simplified by disregarding our experience of other human beings. (Todes 2001: 1; italics in original)

And a reviewer points out:

> the book bypasses entirely the fundamental human experiences of sociality and language – instead one could read Todes thinking that humans are hermits working out the meaning and efficacy of their participation in the world. The kinds of insights later hermeneuts and constructionists offer [these are people working in what I have called the 'Wittgensteinian tradition HMC] – that the categories we use to make our experience know-able and habit-able are accessible human and cultural constructions – were not available to Todes. (Strong 2004: 521)

For an elaborated discussion of these points, see Collins (2016; <http://arxiv.org/abs/1607.08224>).

63 Sacks (2011). Lenat, a computer scientist, runs a project called 'CYC', which aims to incorporate all the knowledge from the world's encyclopedias into what will then be a universally knowledgeable computer. This project seems misplaced because, at best, knowledge is continually evolving, but his remarks on Madeleine seem right.

64 Dreyfusians, failing to spot the difference between individuals and humans as a whole, paint themselves into a corner. Claiming that no one understands a practical

expertise unless they can practise it, they create an impossible world of social isolates – each of us only properly understands the world that we live in – without the possibility of complex division of labour or much else in the way of fruitful social interaction.

65 Collins and Evans (2014); Collins et al. (forthcoming).
66 Wells (1934/1904).

Chapter 5 Six Levels of Artificial Intelligence
67 Russell and Norvig (2003: 48–52).
68 In Collins, Clark and Shrager (2008), I use the term 'culture suckers' for what I refer to here as culture-consumers.
69 Available at: <www.slate.com/id/3650/entry/23905/>.
70 Disappointingly, Deep Mind, the builders of AlphaGo, would not offer a comment on this surmise, nor on my question (see Chapter 10) about whether they could foresee deep learning mastering Winograd schemas.
71 Dennis Hassabis, DeepMind CEO, 'Artificial Intelligence (AI) Invents New Knowledge and Teaches Human New Theories', published on YouTube on 13 April 2017, at 16 minutes 8 seconds.
72 A philosophical way of approaching the question would be to define certain things that humans do as, by definition, only doable by humans. In my 2010 book on *Tacit and Explicit Knowledge* I define tacit knowledge as knowledge that has not or cannot be explicated, and I spend a lot of time coming up with four meanings of 'explicated' so we know what 'not-explicated' means. These four meanings all refer to humans so one way I could analyse what Hassabis says regarding AlphaGo is to say that humans are the locus of explication and therefore the locus of the tacit, and so machines *cannot* possess tacit knowledge or intuition. The alternative,

we could argue, would be that something like sieves have intuition of how to separate large things from small things since they, too, do it without conscious understanding. But this 'philosophical' approach seems unsatisfactory as it appears to answer the question without considering whether a program like AlphaGo has achieved something that other programs have not. The 'social constructivist' way of approaching the question would be to look at the history of these things and note how claims for mechanical reproduction of essential human abilities tend to be made every time that there is a new technological breakthrough, from the telephone exchange onwards. A nice example is the title of a much publicized book from as long ago as 1979, called *Machines Who Think* (McCorduck 1979). We could say, then, that there is nothing more to the accusation of anthropomorphism than the way people do or don't apply it, and this is simply a historical and social variable. The analysis would be interesting but still not tell us whether AlphaGo was using intuition.

73 Collins and Kusch (1998).

74 And it could be that Chess and Go could change to include points for style and come to be scored in the way, say, ice-dancing is scored. What counts as good style would then become a matter of the social and we would no longer be talking of technically defined goals.

75 Interactional expertise would become archaic if not continually refreshed by interaction with the practising community (Collins 2017: Ch. 14, or <http://arxiv.org/abs/1607.07373>).

76 Collins and Kusch (1998: 39).

77 This does not mean that interactional expertise is fragile: the managers of technical projects make large and

serious contributions to enterprises, the technical culture of which they learn without practising, but they too are not the creators of these enterprises.

78 We did a small experiment on the difference between speech and text. We found that when a passage of speech taken from an interview with a scientist that had a liberal scattering of hesitancies – 'ums and ehrs' – was transcribed so as to include the hesitancies, it gave the impression of far less certainty when read than when spoken. The results were striking. Removing the hesitancies from the transcript led to the level of certainty being similar to that of the spoken passage (<http://arxiv.org/abs/1609.01207>). Any program gathering information from text and speech will have to take this kind of thing into account. Hibbard (<https://arxiv.org/abs/1411.1373>) thinks that computers could become socialized by monitoring our interactions with the electronic assistants that we will all use some time in the future, but they will have to know that speech and text are rather different.

79 Likewise, imagine that all the ambitions of the creators of deep learning were to be achieved (irrespective of the arguments presented here), it still wouldn't be Level IV. The training sets for deep learning are so huge that, as Geoffrey Hinton pointed out to me, humans cannot learn this way or they would be presented with multiple new images to learn from – cars, cheetahs, giraffes and so on – every second of their lives. Indeed, Hinton, it seems, has lost confidence in deep learning because it requires so many labelled images to learn from, whereas ordinary humans do not, and he has said that AI needs to start all over again: <https://www.axios.com/ai-pioneer-advocates-starting-over-2485537027.html>. Yoshua Bengio likewise explained to me that a remaining challenge was

to work out how to train deep-learning programs on as few examples as humans learn from. Currently, deep learning works differently from the way humans work, even if we allow that it achieves the same level of success, so we are looking at Level III not Level IV at best.

80 There is nothing incompatible here with what I will call 'the modified sociological ideology' because it involves top-down pattern recognition of phrases used within a particular culture. To do the experiment with individual phrases properly one should get someone else to invent a new garbled passage and break it up – readers of this book are already too familiar with the 'Cmabrigde Uinervtisy' passage.

Chapter 6 Deep Learning: Precedent-Based, Pattern-Recognizing Computers

81 More and more, however, programmers are expending the huge effort because machines using these technologies may be making decisions in life-threatening situations and their behaviour may have to be explained in courtrooms and the like .

82 In the terminology of the Periodic Table of Expertises (Collins and Evans 2007) this explanation is only a little better than 'beer mat knowledge', but it is just better enough to show that in principle a machine can learn from experience.

83 I am delighted that one of the technically proficient readers of the manuscript told me my account was too accurate to be referred to as a 'just so' story. But for a more informative account that is still written in accessible language, see: <https://www.technologyreview.com/s/608911/is-ai-riding-a-one-trick-pony/>.

84 The typically mistaken account can be found in Kurzweil (2012: 18 ff.). A more accurate and very simple account

of the experiment and subsequent developments can be found in Collins and Pinch (1993/8: Ch. 2).

85 In addition, Ernest Davis remarks: 'As far as I can judge, what BACON did bore no relation whatever to the problem that Kepler faced. BACON was given a table of year lengths and orbital radii, and succeeded in fitting them to the curve $R^3 = cT^2$. Kepler started with the positions of the planets in the sky as seen from earth, with no depth information at all, and had to figure out where they were in space. It is an incomparably harder problem.'

86 I learned only later (thanks to Hector Levesque) that, some decades back, 'bottom up' and 'top down' were standard usage in debates about computer vision. My use of the terms has nothing to do with this, but I suspect the similarity of terms in both cases is far from coincidental: the structure of the two kinds of problems is similar – in fact, in some ways they are the same problem.

87 This modified model was initially used to analyse the relationship between tacit and explicit knowledge in Collins (2010).

88 The term 'affordance' doesn't *explain* anything but is useful for *describing* how things relate.

89 Collins and Reber (2013).

90 Though this must depend on how demanding the examples are.

91 Ernest Davis tells me that other leading figures in the field of computer vision are more engaged with the need to consider top-down constraints than Hinton.

92 Thanks to Yoshua Bengio for clarification on this point. He (private communication) affirms that the term 'implicit supervision' is a good one. His preferred description is:

One trivial example is when the computer simply observes the utterances and writings produced by humans (and that we do already, of course). A more advanced case is when the computer dialogues with humans and uses that experience to learn about language (and that kind of data is already being accumulated by companies like Apple and Google from our phones, but up to now this is not something we can do well, meaning that not much more than pure observation is deduced by the learning machine, whereas we could easily imagine that dialogue could be used to actually extract much more information from humans, without requiring those humans to serve as actual teachers).

93 Blackwell (2017: 7–8).

94 One might prefer this culture to the others that are currently available but one should know it is a choice, not an inevitability. Maybe the 'Search for Extraterrestrial Life' has been a failure because all the advanced civilizations on other planets made different choices! The author, by the way, does prefer this culture to anything else we currently have on Earth: see his 2014 *Are We All Scientific Experts Now?* And see also his 2017, co-authored, *Why Democracies Need Science*, but we can imagine better options, surely!

Chapter 7 *Kurzweil's Brain and the Sociology of Knowledge*

95 See Collins (2011b) for the 'fractal model' which described the relationship between large and small social groups; and see Chapter 8 below.

96 See, for example, <https://www.change.org/p/prof-karsten-danzmann-beantworten-sie-bitte-3-fragen-%C3%BCber-das-ligo-experiment>). Non-German speakers will find an English translation below the German.

97 Less exciting is that Bengio replaces what sociologists call culture with the high priest of evolution, Richard

Dawkins's, notion of 'memes'. When I queried this, he wrote:

> As a computer scientist and a pioneer of deep learning, I see a very strong and important analogy between the notion of 'nugget of knowledge' (a meme, an idea), which can be composed with others to form solutions to problems, or new memes, and the notion of genes and genetic evolution. (Bengio, private communication, 2 September 2017)

Sociologists find the notion of 'memes' strangely redundant, though it is understandable that those who work with artificial intelligence would be attracted to an idea that seems to reduce sociology to evolution. For the sociologist the problem with memes is that, insofar as they are supposed to be analogous with genes, one would expect to find something equivalent to a well-defined physical environment that would supply the conditions for the survival of some memes rather than others. But the environment within which one idea survives and another dies (it might be witches vs mortgages) is culture itself, so talk of memes returns us immediately to the sociological starting point even as it provides, for some, the comforting sense that a kind of superior and more objective science has replaced the sociology.

98 For a discussion of solutions to the problem for science, see Collins, Bartlett and Reyes-Galindo (2017).

Chapter 8 How Humans Learn What Computers Can't
99 See Collins and Evans (2007).
100 The immediate problem for computers will be discussed again in Chapter 9 when we look at Google's 'Pagerank'.
101 See Collins (2017), *Gravity's Kiss*, for a real-time account of this discovery and its confirmation and acceptance. *Gravity's Shadow* (Collins 2004a) details the earlier

sociological history of the field, while *Gravity's Ghost and Big Dog* details two data-analysis rehearsals prior to the discovery, explaining some of the technical details in great detail.

102 See note 50 in Chapter 4.

103 This was under our protocol where judges have to supply a confidence level from 1–4 to accompany their judgement and we count level 1 and 2 as 'don't know'. If we ignore this restriction, the voting was: seven in favour of Collins, one correct choice and one 'don't know'.

104 Descriptions of the formation of trust among politicians of the Soviet era often appear to involve the consumption of large quantities of vodka, and trust among modern American businessmen seems often to involve athletic pursuits such as mountain-biking, squash, tennis or golf.

105 There is also some guarding of property: see Collins's (2017), *Gravity's Kiss*. There is a large literature on 'boundary work', which follows from the sociology of scientific knowledge. A key concept is the 'core-set' – the very small group of specialists that does the real work at the centre of a scientific controversy (Collins 1985/92). Creating, understanding and policing the boundaries of physics is a really difficult problem, having as much to do with sociological understanding as science per se, but still crucial in the making of science (Collins, Bartlett and Reyes-Galindo 2017).

106 My 2004a book, *Gravity's Shadow*, and a commentary on it found in Collins and Evans (2015), details me having quite heated technical arguments with the then Director of the Laser Interferometer Gravitational-Wave Observatory: I could only have done this in face-to-face interaction and given quite a long history of previous face-to-face interaction.

107 I won't try to explain little dogs here. They are discussed

in Collins (2017) and at greater length in Collins (2013a) and briefly in Appendix 2 of this book.

108 The Weber paper is Weber and Radak (1996); the test with arXiv's automatic filter is reported in Collins, Ginsparg and Reyes-Galindo (2016). Some readers will be thinking 'doesn't the discovery in August 2017 that the gravitational waves from an inspiraling binary neutron star system were coincident with a gamma ray burst prove that Weber was right after all?' If they are thinking that, it is a nice illustration of the insider/outsider point. Weber's gravitational-wave detector was many orders of magnitude too insensitive to see such systems.

109 Contrast the model espoused by Dreyfus and Todes.

Chapter 9 Two Models of Artificial Intelligence and the Way Forward

110 Levesque (2017) has a nice version of this kind of argument in respect of the Chinese Room (Chapter 3). He imagines that the Chinese speaker in the Chinese Room learned Chinese by memorizing just the kind of look-up tables that Searle posits as the resource used by the non-Chinese speaker; in other words, the brain *is* a Chinese Room. This also shows how hard it is to pin down internal states.

111 Weizenbaum (1976), Winograd and Flores (1986); Levesque's and Davis's contributions are mentioned throughout.

112 For the 'past-sell-by-date' criterion for scientific policy-making, see Collins and Weinel (2011) and Collins, Bartlett and Reyes-Galindo (2017).

113 See Collins and Evans (2017: Ch. 2) for an analysis of the norms of science.

Chapter 10 The Editing Test and Other New Versions of the Turing Test

114 It arises out of work by Hector Levesque, for example, Levesque (2017) and, earlier, Levesque et al. (2012). The new competition reflects renewed interest in the Turing Test marked by the special issue of *AI Magazine* 37/1 (spring 2016). Mostly, the articles in the special issue concern extensions of the Turing Test to take into account bodily activities or the like, but there were some papers that stick with language as the key. For a summary, see Marcus, Rossi and Veloso (2016).

115 Collins and Evans (2007). As noted in Chapter 9, each specialism in a society has its own ubiquitous expertise related in a fractal-like way – e.g., in gravitational-wave physics the ubiquitous expertise includes the general theory of relativity and the second law of thermodynamics, whereas for the average citizen, ubiquitous expertise is to do with living life in general.

116 In Levesque (2017: 59), in reference to his version of the test, we find: 'No expert judges are needed.'

117 And set up by Morgenstern, Davis and Ortiz; see their 2016 and also <http://commonsensereasoning.org/wino grad.html>.

118 Private communication from Ernest Davis, 27 August 2016. This report was subsequently published in *AI Magazine*: Morgenstern, Davis and Ortiz (2016).

119 It can be found on page 11. Thanks to Anthony Cohn for pointing this out to me. Winograd's thesis is entitled 'Procedures as a Representation for Data in a Computer Program for Understanding Natural Language', Massachusetts Institute of Technology, 1971.

120 Incidentally, if we try:

> The council women refused to give out a permit for a demonstration because they feared violence

We get the correctly gendered:

> Les femmes du conseil ont refusé de donner un permis pour une manifestation parce qu'elles craignaient la violence

But if we try:

> The council women refused to give the men a permit for a demonstration because they feared violence

We get:

> Les femmes du conseil ont refusé de donner aux hommes un permis de manifestation parce qu'ils craignaient la violence

So it looks like the masculine is the default wherever there is ambiguity. This is stuff you can try yourself.

121 Davis (2016).
122 You can try these questions out on Siri!
123 Collins (2010) divides tacit knowledge into three kinds: relational, somatic and collective.
124 For example, Papert (1968), which is aggressively entitled 'The Artificial Intelligence of Hubert Dreyfus: A Budget of Fallacies' (e.g., section 1.4).
125 Private communications, spring/summer 2017.
126 Collins (2013b).
127 Interestingly, in the second half of Davis (2016), some tests are developed which go a little way up the Y-axis, as they demand high-school science comprehension.
128 Thanks to Yoshua Bengio for reminding me of word embedding as a solution.
129 '1.000' would mean that the words always appear together.
130 As Ernie Davis pointed out to me, 'juggle' doesn't appear either but you can juggle a watermelon (just). The power, if it is there, will come from the relationship between all the words in the corpus, not just the

relationship between one or two; here we merely get a sense of the potential.

131 In *Gravity's Shadow* I write something that can be thought about in the same way:

> The less literal side to truth making is still more interesting. Conferences are the places where the community learns the *etiquette* of today's truth; it learns what words and usages are properly uttered in polite company. Thus, in conference after conference, Joe Weber [a pioneering GW scientist whose claims to have seen high fluxes of gravitational wave were discredited by 1975] would stand up and present his papers, explaining that he had found gravity waves long ago, and the delegates learned that the right response was to quietly move on to the next paper. And later, conferences would happen without the physical presence of Joe Weber or even his virtual presence in the vibrations of the airwaves that constitute words. In my first day at the [1996] Pisa [GW] conference, during which I listened to every paper, Weber's name was mentioned just once, in passing (pp. 451–2).

132 See: (<http://www.cs.nyu.edu/faculty/davise/papers/WinogradSchemas/WSCollection.xml>).

133 Though Davis tells me it is easy to construct new ones, he admits he does not know how large the list of potential Winograd schemas is.

134 See: <https://en.wikipedia.org/wiki/Cyc>. Bengio's remark (above, pp. 181–2) about using videogames to teach deep-learning computers reminds us that there is a lot of commonsense knowledge already programmed into the non-fantastic versions of these.

135 Crowd-sourcing as a possibility is mentioned in Davis (2017).

136 Davis (private communication, 31 August 2017) says that whether the list of common-sense questions is intractably large is not known, but he thinks not. For more discussion of this kind of knowledge, see Perlis, (2016).

137 Bengio (private communication, 31 August 2017) says that he believes deep learning will be able to solve these problems, 'By doing exactly what you say, i.e., by deeply embedding a learning machine in the bath of language, and furthermore by doing it in what we call a grounded way, i.e., in relation to some environment where the words refer to things, places, agents, actions, etc. in this environment.'

I cannot get my head round this solution without imagining the computer being something so like a socialized human as to be a socialized human being. But I am not trying to prophesy, just explain where we currently are, and, should it turn out that Bengio is right, show what novel things must have been accomplished to make these fully fluent computers possible; it does seem that both sides agree that it will involve embedding in the language bath of social life.

138 See: <http://www.cs.nyu.edu/faculty/davise/papers/Win ogradSchemas/ WSCollection.xml>.

Appendix 1 How the Internet Works Today

139 Remember that brute strength, as we have seen from the experiment with the simple anagram-finding program, may be enough to do all kinds of repairs without relying on Google. Incidentally, there is a ready-made program that will jumble words for you (<http://www. bluestwave.com/toolbox_letter_scrambler.php>), but I recommend doing it yourself because it would be all too easy for someone to pick up when that program has been accessed and reverse what it does to make sense of things.

References

Barrow, John D. 1999. *Impossibility: The Limits of Science and the Science of Limits.* New York: Oxford University Press.

Bengio, Yoshua 2012. 'Evolving Culture vs Local Minima', 29 November 2012. Available at: <http://arXiv:1203.2990v2> [cs.LG].

Bengio, Yoshua 2014. 'Deep Learning and Cultural Evolution', in *Proceedings of the Companion Publication of the 2014 Annual Conference on Genetic and Evolutionary Computation*, pp. 1–2. Available at: <http://dl.acm.org/citation.cfm?id=2598395>.

Biederman, Irving 1987. 'Recognition-by-Components: A Theory of Human Image Understanding', *Psychological Review* 94: 115–47.

Blackwell, Alan F. 2015. 'Interacting with an Inferred World: The Challenge of Machine Learning for Humane Computer Interaction', in *Proceedings of Critical Alternatives: The 5th Decennial Aarhus Conference*, pp. 169–80. Available at: <http://dx.doi.org/10.7146/aahcc.v1i1.21197>.

Blackwell, Alan F. 2017. 'Objective Functions, Deep Learning and Random Forests', Contribution to *Science in the Forest, Science in the Past*, Needham Institute, Cambridge. Available at: <http://www.cl.cam.ac.uk/~afb21/publica tions/Blackwell-ObjectiveFunctions.pdf>.

Boden, Margaret A. 2008. *Mind as Machine: A History of Cognitive Science*. Oxford: Clarendon Press.

Collins, Harry 1981. 'What is TRASP: The Radical Programme as a Methodological Imperative', *Philosophy of the Social Sciences* 11: 215–24.

Collins, Harry 1985/92. *Changing Order: Replication and Induction in Scientific Practice*. Beverley Hills and London: Sage; 2nd edition 1992, Chicago: University of Chicago Press.

Collins, Harry 1990. *Artificial Experts: Social Knowledge and Intelligent Machines*. Cambridge: MIT Press.

Collins, Harry 1996. 'Embedded or Embodied? A Review of Hubert Dreyfus's *What Computers Still Can't Do*', *Artificial Intelligence* 80/1: 99–117.

Collins, Harry 2000. 'Four Kinds of Knowledge, Two (or maybe Three) Kinds of Embodiment, and the Question of Artifical Intelligence', in Jeff Malpas and Mark A. Wrathall (eds), *Heidegger, Coping, and Cognitive Science: Essays in Honor of Hubert L. Dreyfus*, vol. 2. Cambridge: MIT Press, pp. 179–95.

Collins, Harry 2001. 'Tacit Knowledge, Trust, and the Q of Sapphire', *Social Studies of Science* 31/1: 71–85.

Collins, Harry 2004a. *Gravity's Shadow: The Search for Gravitational Waves*. Chicago: University of Chicago Press.

Collins, Harry 2004b. 'How Do You Know You've Alternated?', *Social Studies of Science* 34/1: 103–6.

Collins, Harry 2007. 'Mathematical Understanding and the Physical Sciences', in Collins (ed.), *Case Studies of Expertise and Experience: Special issue of Studies in History and Philosophy of Science* 38/4: 667–85.

Collins, Harry 2010. *Tacit and Explicit Knowledge*. Chicago: University of Chicago Press.

Collins, Harry 2011a. *Gravity's Ghost: Scientific Discovery in the Twenty-First Century*. Chicago: University of Chicago Press.

Collins, Harry 2011b. 'Language and Practice', *Social Studies of Science* 41/2: 271–300.

Collins, Harry 2013a. *Gravity's Ghost and Big Dog: Scientific Discovery and Social Analysis in the Twenty-First Century*. Chicago: University of Chicago Press (includes a reprint of Gravity's Ghost).

Collins, Harry 2013b. 'Three Dimensions of Expertise', *Phenomenology and the Cognitive Sciences* 12/2: 253–73.

Collins, Harry 2014. *Are We All Scientific Experts Now?* Cambridge: Polity.

Collins, Harry 2016. 'Interactional Expertise and Embodiment', in Jorgen Sandberg, Linda Rouleau, Ann Langley and Haridimos Tsoukas (eds), *Skilful Performance: Enacting Expertise, Competence, and Capabilities in Organizations: Perspectives on Process Organization Studies (P-PROS)*, vol. 7. Oxford: Oxford University Press. Available at: <http://arxiv.org/abs/1607.08224>.

Collins, Harry 2017. *Gravity's Kiss: The Detection of Gravitational Waves*. Cambridge: MIT Press. For Chapter 14, see: <http://arxiv.org/abs/1607.07373>.

Collins, Harry Forthcoming. *Forms of Life: The Method and Meaning of Sociology*. Cambridge, MA: MIT Press.

Collins, Harry and Evans, Robert 2007. *Rethinking Expertise*. Chicago: University of Chicago Press.

Collins, Harry and Evans, Robert 2014. 'Quantifying the Tacit: The Imitation Game and Social Fluency', *Sociology* 48/1: 3–19.

Collins, Harry and Evans, Robert 2015. 'Expertise Revisited I – Interactional expertise', *Studies in History and Philosophy of Science* 54: 113–23; preprint at: <http://arxiv.org/abs/1611.04423>.

Collins, Harry and Evans, Robert 2017. *Why Democracies Need Science*. Cambridge: Polity.

Collins, Harry and Kusch, Martin 1998. *The Shape of Actions: What Humans and Machines Can Do*. Cambridge: MIT Press.

Collins, Harry and Pinch, Trevor 1993/1998. *The Golem: What Everyone Should Know About Science*. Cambridge and New York: Cambridge University Press; New editions, 1998 (sub-titled *What You Should Know about Science*), reissued as Canto Classic in 2012.

Collins, Harry and Pinch, Trevor 1995. *Dr Golem: How to Think about Medicine*. Chicago: University of Chicago Press.

Collins, Harry and Pinch, Trevor 2005. *Dr Golem: How to Think about Medicine*. Chicago: University of Chicago Press.

Collins, Harry and Reber, Arthur 2013. 'Ships that Pass in the Night', *Philosophia Scientiae* 17/3: 135–54.

Collins, Harry and Weinel, Martin 2011. 'Transmuted Expertise: How Technical Non-experts Can Assess Experts and Expertise', *Argumentation: Special Issue on Rethinking Arguments from Experts* 25/3: 401–13.

Collins, Harry, Bartlett, Andrew and Reyes-Galindo, Luis 2017. 'The Ecology of Fringe Science and its Bearing on Policy', *Perspectives on Science* 25/4: 411–38. Available at: <http://arxiv.org/abs/1606.05786>. (An earlier version promulgated as 'The Ecology of Fringe Science and its Bearing on Policy' is available at: <http://arxiv.org/abs/1606.05786>.)

Collins, Harry, Clark, Andy and Shrager, Jeff 2008. 'Keeping the Collectivity in Mind?', *Phenomenology and the Cognitive Sciences* 7/3: 353–74.

Collins, Harry, Ginsparg, Paul and Reyes-Galindo, Luis 2016. 'A Note Concerning Primary Source Knowledge', *Journal of the Association for Information Science and Technology*, May DOI: 10.1002/asi. Available at: <http://arxiv.org/abs/1605.07228>.

Collins, Harry, Green, R. H. and Draper, R. C. 1985. 'Where's the Expertise: Expert Systems as a Medium of Knowledge Transfer', in M. J. Merry (ed.), *Expert Systems 85*, Cambridge: Cambridge University Press, pp. 323–4.

Collins, Harry, Evans, Robert, Pineda, Sergio and Weinel, Martin 2016. 'Modelling Architecture in the World of Expertise', *Room One Thousand* 4: 23–34.

Collins, Harry, Hall, Martin, Evans, Robert and O'Mahony, Hannah Forthcoming. *Imitation Games: A New Method for Investigating Societies*. Cambridge, MA: MIT Press.

Davis, Ernest 2016. 'How to Write Science Questions that are Easy for People and Hard for Computers', *AI Magazine* 31/1: 13–22.

Davis, Ernest 2017. 'Logical Formalizations of Commonsense Reasoning: A Survey', *Journal of Artificial Intelligence Research* 59: 651–723.

Deacon, Terrence, W. 1997. *The Symbolic Species: The Co-evolution of Language and the Brain*. New York: Norton.

Dreyfus, Hubert 1967. 'Why Computers Must Have Bodies in Order to Be Intelligent', *The Review of Metaphysics* 21/1: 13–32.

Dreyfus, Hubert L. 1992 [1972]. *What Computers Can't Do*. Cambridge, MA: MIT Press.

Fleck, Ludwik 1979. *Genesis and Development of a Scientific Fact*. Chicago: University of Chicago Press (first published in German in 1935 as *Entstehung und Entwicklung einer wissenschaftlichen Tatsache: Einführung in die Lehre vom Denkstil und Denkkollektiv*).

Giles, Jim 2006. 'Sociologist Fools Physics Judges', *Nature* 442: 8.

Kuhn, Thomas S. 1962. *The Structure of Scientific Revolutions*. Chicago: University of Chicago Press.

Kurzweil, Ray 2005. *The Singularity is Near: When Humans Transcend Biology*, New York: Viking Penguin.

Kurzweil, Ray 2012. *How to Create a Mind: The Secret of Human Thought Revealed*. New York: Viking Penguin.

Laland, Kevin, N. 2017. *Darwin's Unfinished Symphony: How Culture Made the Human Mind*. Princeton NJ: Princeton University Press.

Levesque, Hector Unpublished. 'The Winograd Schema Challenge'. Available at: <http://www.cs.toronto.edu/~hec tor/Papers/winograd.pdf>.

Levesque, Hector 2014. 'On Our Best Behaviour', *Artificial Intelligence* 212: 27–35.

Levesque, Hector 2017. *Common Sense, the Turing Test, and the Quest for Real AI*. Cambridge, MA.: MIT Press.

Levesque, Hector, Davis, Ernest and Morgenstern, Leora 2012. *The Winograd Schema Challenge*. Proceedings of Principles of Knowledge Representation and Reasoning.

McCorduck, Pamela 1979. *Machines Who Think: A Personal Inquiry into the History and Prospects of Artificial Intelligence*. San Francisco: W. H. Freeman.

Marcus, Gary, Ross, Francesca and Veloso, Manuela 2016. 'Beyond the Turing Test', *AI Magazine* 37/1: 3–4.

Morgenstern, Leora, Davis, Ernest and Ortiz, Charles, L. 2016. 'Planning, Executing, and Evaluating the Winograd Schema Challenge', *AI Magazine* 37/1: 50–4.

O'Neil, Cathy 2017. *Weapons of Math Destruction*. Harmondsworth: Broadway Books.

Papert, Seymour 1968. 'The Artificial Intelligence of Hubert

Dreyfus: A Budget of Fallacies', MIT Artificial Intelligence Memo No. 154. Available at: <https://dspace.mit.edu/handle/1721.1/6084>.

Perlis, Don 2016. 'Five Dimensions of Reasoning in the Wild, *AAAI-16 Proceedings of the Thirtieth AAAI Conference on Artificial Intelligence*. Phoenix, Arizona, February 12–17, 2016: 4152–56.

Perlis, Donald, Purang, Khemdut and Andersen, Carl 1998. 'Conversational Adequacy; Mistakes are the Essence', *International Journal of Human–Computer Studies* 48: 553–75.

Russell, Stuart J. and Norvig, Peter 2003. *Artificial Intelligence: A Modern Approach* (2nd edn). Upper Saddle River, New Jersey: Prentice Hall.

Sacks, Oliver 2011. *The Man Who Mistook His Wife for a Hat*. London: Picador.

Selinger, Evan 2003. 'The Necessity of Embodiment: The Dreyfus–Collins Debate', *Philosophy Today* 47/3: 266–79.

Selinger, Evan, Dreyfus, Hubert and Collins, Harry 2007. 'Embodiment and Interactional Expertise', *Studies in History and Philosophy of Science* 38/4: 722–40.

Strong, Tom 2004. 'Bodies and Thinking Motion', *Janus Head* 7/2: 516–22. Available at: <http://www.janushead.org/7-2/Todes.pdf>.

Suchman, L. A. 1987. *Plans and Situated Action: The Problem of Human–Machine Interaction*. Cambridge: Cambridge University Press.

Tegmark, Max 2017. *Life 3.0: Being Human in the Age of Artificial Intelligence*. New York: Alfred Knopf.

Todes, Samuel 2001. *Body and World*. Cambridge, MA: MIT Press.

Turing, Allan. M. 1950. 'Computing Machinery and Intelligence', *Mind* LIX 236: 433–60.

Vinge, Vernor 1993. 'The Coming Technological Singularity:

How to Survive in the Post-Human Era', *VISION-21 Symposium*, sponsored by NASA Lewis Research Center and the Ohio Aerospace Institute, March 30–31.

Weber, Joseph and Radak, B. 1996. 'Search for Correlations of Gamma-Ray Bursts with Gravitational-Radiation Antenna Pulses', *Il Nuovo Cimento B Series 11* 111/6: 687–92.

Weizenbaum, Joseph 1976. *Computer Power and Human Reason: From Judgement to Calculation*. San Francisco: W. H. Freeman.

Wells, H. G. 1998 [1904]. 'The Country of the Blind', repr. in *The Complete Short Stories of H. G. Wells*, ed. John Hammond. London: Phoenix Press, pp. 846–70.

Winch, Peter. G. 1958. *The Idea of a Social Science*. London: Routledge and Kegan Paul.

Winograd, T. and Flores, F. 1986. *Understanding Computers and Cognition: A New Foundation for Design*. New Jersey: Ablex.

Wittgenstein, Ludwig 1953. *Philosophical Investigations*. Oxford: Blackwell.

Xin Luna Dong, Evgeniy Gabrilovich, Kevin Murphy, Van Dang Wilko Horn, Camillo Lugaresi, Shaohua Sun, Wei Zhang 2015. 'Knowledge-Based Trust: Estimating the Trustworthiness of Web Sources'. Available at: <http://arxiv.org/pdf/1502.03519v1.pdf>.

Index